CULTURE, PLACE, AND NATURE

STUDIES IN ANTHROPOLOGY AND ENVIRONMENT

K. Sivaramakrishnan, Series Editor

Centered in anthropology, the Culture, Place, and Nature series encompasses new interdisciplinary social science research on environmental issues, focusing on the intersection of culture, ecology, and politics in global, national, and local contexts. Contributors to the series view environmental knowledge and issues from the multiple and often conflicting perspectives of various cultural systems.

CULTURE, PLACE, AND NATURE

STUDIES IN ANTHROPOLOGY AND ENVIRONMENT

The Kuhls of Kangra: Community-Managed Irrigation in the Western Himalaya, by Mark Baker

The Earth's Blanket: Traditional Teachings for Sustainable Living, by Nancy Turner

Property and Politics in Sabah, Malaysia: Native Struggles over Land Rights, by Amity A. Doolittle

Border Landscapes: The Politics of Akha Land Use in China and Thailand, by Janet C. Sturgeon

From Enslavement to Environmentalism: Politics on a Southern African Frontier, by David McDermott Hughes

Ecological Nationalisms: Nature, Livelihood, and Identities in South Asia, edited by Gunnel Cederlöf and K. Sivaramakrishnan

Tropics and the Traveling Gaze: India, Landscape, and Science, 1800–1856, by David Arnold

Being and Place among the Tlingit, by Thomas F. Thornton

Forest Guardians, Forest Destroyers: The Politics of Environmental Knowledge in Northern Thailand, by Tim Forsyth and Andrew Walker

Nature Protests: The End of Ecology in Slovakia, by Edward Snajdr

Wild Sardinia: Indigeneity and the Global Dreamtimes of Environmentalism, by Tracey Heatherington

Tahiti beyond the Postcard: Power, Place, and Everyday Life, by Miriam Kahn

Forests of Identity: Society, Ethnicity, and Stereotypes in the Congo River Basin, by Stephanie Rupp

Enclosed: Conservation, Cattle, and Commerce among the Q'eqchi' Maya Lowlanders, by Liza Grandia

Puer Tea: Ancient Caravans and Urban Chic, by Jinghong Zhang

PUER TEA

Ancient Caravans and Urban Chic

JINGHONG ZHANG

A China Program Book

UNIVERSITY OF WASHINGTON PRESS

Seattle & London

This book was supported in part by the China Studies Program,
a division of the Henry M. Jackson School of International Studies
at the University of Washington.

Printed and bound in the United States of America
Composed in Warnock Pro, a typeface designed by Robert Slimbach
22 21 20 19 18   7 6 5 4 3

UNIVERSITY OF WASHINGTON PRESS
www.washington.edu/uwpress

LIBRARY OF CONGRESS CATALOGING-IN-PUBLICATION DATA
Zhang, Jinghong.
Puer tea : ancient caravans and urban chic / Jinghong Zhang.
pages cm
(Culture, place, and nature)
"A China Program Book."
Includes bibliographical references and index.
ISBN 978-0-295-99322-5 (hb : alk. paper)
ISBN 978-0-295-99323-2 (pb : alk. paper)
1. Tea—China—Yunnan Sheng.
2. Tea—Social aspects—China—Yunnan Sheng.
3. Tea trade—China—Yunnan Sheng.
I. Title.
GT2907.C6Z4366 2013
394.1'5095135—dc23 2013027326

The paper used in this publication is acid-free and meets the minimum
requirements of American National Standard for Information Sciences—
Permanence of Paper for Printed Library Materials, ANSI Z39.48 1984. 8

For my maternal grandma and my mum

CONTENTS

FOREWORD

In this fascinating study of Puer tea and the life of rural Chinese and others in Yunnan, we can see many transformations, subtle and dramatic, at work. Jinghong Zhang hails from Yunnan and returns there repeatedly to conduct fieldwork that traces the relationship of Puer tea to the landscape, livelihoods, and tastes in newly prosperous China in the early twenty-first century. The study is informed by deep ethnography as well as by admirable plant knowledge, nimble literary imagination, and thorough acquaintance with social history, regional geography, and aesthetics. Film and photographs enhance the research and presentation, as the author enriches environmental anthropology with humanistic inquiry that is unusually adept at turning a literary trope into a tool of social analysis.

The literature on the global trade in agricultural commodities includes several anthropological studies that reveal the merit of closely analyzing how commodity chains come into existence. Some of the more recent studies focus on commodities that appeal to fashions for ethnic foods or sustainably grown or harvested foods, demonstrating how commodity cultures that are prevalent across global markets become opportunities or constraints for local areas and national economies. This study offers such an examination, built on the case of Puer tea from China. It is blessed by the long-term engagement of the author with the tea-growing region in Yunnan Province and her deep knowledge of the tea plant and the making and consumption of Puer tea. Like a good commodity chain researcher, she has followed the tea from fields to traders and consumers, doing research in south China, Hong Kong, and Taiwan along the way.

Zhang's study resonates with others that are beginning to focus on issues of cultural heritage in postreform China and on the ways in which the cultural history of China has been defined in recent decades. Built environments, indigenous peoples, karst landscapes, and tea are all part of the production of a new national cultural heritage in which precommunist history and its material record are being reclaimed in new ways. Zhang places these concerns at the heart of her inquiry by investigating how the consumption of Puer tea is caught up in questions of experiencing authenticity. The adventures and uncertainties brought about by this distinctive tea, a cultivated taste, are told with reference to the multidimensional metaphor evoked by the concept of *jianghu*. As one of the reviewers of the manuscript noted, "The popular and enduring archetype of *jianghu* was the wandering swordsman, an independent worldly-wise character whose exploits revealed the seamy underbelly of society as well as its noblest ideals. Villagers, traders, promoters, spin masters, collectors, and connoisseurs are all *jianghu* characters in the amorphous theater of Puer tea."

Zhang not only takes up the tea and its travels but constructs a vivid account (illuminated further by the excellent use of images and film) of the spaces in which Puer tea is made, traded, and consumed. She wishes that place to be understood as always emergent, a somewhat inchoate and unruly space where rival meanings jostle for supremacy and where hierarchies of value remain unstable. This instability is in part caused by the state's inability to play a defining role. Arguments are illustrated through a variety of activities—from cultivation, certification, and tasting to packaging—which the ethnography carefully unveils, a task in which it is liberally assisted by great photographs. Zhang crafts her theoretical encounters with a light touch, always finding what new light the case of Puer tea has to shed on them, never letting more arcane dimensions cloud the view. Rarely has a study of commodities and culture so felicitously combined perspectives on economy, imagination, travel, and landscape. This book will surely bring fresh interest to a variety of fields from food to commerce, heritage in land, social life of plants, and national self-fashioning in the light of cosmopolitan environmentalism.

K. Sivaramakrishnan
Yale University
March 2013

ACKNOWLEDGMENTS

Without Andrew Walker, this research on Puer tea would still be an unfulfilled dream. I have benefited greatly from his wisdom, his pragmatic views, and his ability to come straight to the main research issue. He taught me to use specific methods at specific stages, directing me step by step to select, sort, process, and finally brew this tea monograph.

The ethnographic filmmaking of David MacDougall and Judith MacDougall has inspired me to use video as an important component of this research. The planning, filming, and editing for the video companion to this book have benefited enormously from Judith's profound insights, meticulous inspection, and continual kind support. I have learned so much from her about both study and life, from film to food. And her motherly considerations kept me warm through Canberra's winter.

Nicholas Tapp provided detailed, sincere, and prompt advice. His wide knowledge has enriched my reading and use of literature in this book. Many of his comments have kept me thinking and will be useful starting points for future research. Amrih Widodo encouraged me to think about questions from alternative perspectives and in unconstrained ways. I often felt that if I had one-third of his intelligence, the book I produced would have been three times as good. Duncan Campbell kindly assisted me in deepening my understanding of Chinese cultural consumption. Several books he recommended became favorites that I have used in the book.

I completed most of my writing for this book at the Resource Management in Asia-Pacific Program (RMAP) at the Australian National University's College of Asia and the Pacific. I am thankful to RMAP for providing me the environment to undertake this challenging work. I have also benefited from writing groups at the Australian National University organized by Andrew Kipnis and Craig Reynolds, and have collected many useful suggestions from the participants. Pip Deveson helped me convert the films into DVD. She displayed great patience and taught me how to solve technical problems via trial and error. Ann Maxwell Hill, Magnus Fiskesjö, C. Patterson Giersch, and three readers from the University of Washington Press provided encouraging and valuable feedback on the initial manuscript. Lorri Hagman, Tim Zimmermann, Emily Park, Jacqueline Volin, and David Peattie offered detailed editorial suggestions and helped me move along toward publication.

I am grateful to Mu Jihong, Zhang Yi, Zheng Bineng, and Li Jinwen, who guided me into the world of Puer tea. Eddie Tsang, Matthew Chen, Lei Bin, Peter D'Abbs, Gao Fachang, Sidney Cheung, Yu Shuenn-Der, Yang Haichao, Chen Gang, Austin Hodge, Zhu Ping, Zhang Yingpei, Mette Trier, Robin Fash-Boyle, and Søren M. Chr. Bisgaard gave me kind assistance in fieldwork and exchanged excellent insights with me. The experiences of drinking tea and talking with various tea friends have made me feel that my research is full of Puer tea's vitality.

I am lucky to have friendships with Li Chyi-Chang, Masayuki Nishida, Keri Mills, Jakkrit Sangkhamenee, Guan Jia, Li Yinan, Gu Jie, Jin Hao, Lei Junran, Sophie Mcintyre, Rachel P. Lorenzen, Natasha Fijn, Yasir Alimi, and Darja Hoenigman at the Australian National University. They have made life in Canberra more meaningful to me. Mother and Father Chuang kindly allowed me to live at their place for most of my time in Canberra.

It was from Yang Kun that I learned the value of conducting research out of my authentic interest. He contributed many brilliant ideas, some of which influenced the way I have used film. I am very sad that he could not see my final work, as he passed away in March 2010.

Most of all I owe this research to my husband, Jingfeng, who shared many pleasures and difficulties with me throughout my study, and also my parents, brother, and sister, whose belief in me makes any achievement possible. It's far beyond my textual ability to thank them.

My family and friends jokingly call me a "tea doctor" (*chaboshi*). In Chinese, this refers to one who works in a teahouse and serves water for guests to make tea. I would happily be called this if it meant I could express my sincere thanks to all the above people by serving them a cup of tea.

This book draws on some previously published material. Parts of chapters 1, 2, and 5 are based on a 2010 article in the *Journal of Chinese Dietary Culture* 6 (2). Chapter 5 also draws on a 2012 article in *Collected Papers on the Ancient Tea-Horse Road* (Vol. 2). Chapter 8 is a revised version of a 2010 article in the *Journal of Chinese Dietary Culture* 6 (1). Chapter 4 is based on a 2012 article in *Harvard Asia Quarterly* spring/summer XIV (1 and 2).

TRANSLITERATION, NAMES, AND MEASURES

In this book, all transcriptions of standard Chinese (Mandarin), including the Yunnan dialects, follow the standard Pinyin romanization system. I have made exceptions for commonly used names, such as Hong Kong; Cantonese terms that have been accepted in English, such as *kungfu*, *yum cha*, *dim sum*; and other names that have been published elsewhere using Wade-Giles or other romanization systems.

Some personal names are used with permission, such as Zhang Yingpei and Gao Fachang. For others, such as Mr. Zheng, Mr. Li, and Zongming, I have used partially real names, mainly for the convenience of reading in English. For those who were unwilling to have their real names mentioned, I have used pseudonyms. In full Chinese names, consistent with Chinese custom, the surname precedes the given name. In Chinese literature citations, full Chinese names differentiate authors who share the same surname and also the same publishing year.

References to monetary values are mostly in Chinese yuan (¥). Throughout the time of my fieldwork in 2007, US$1 was approximately equivalent to ¥7. In one case I used Singapore dollars. S$1 at that time was approximately equivalent to US$0.57 (¥4).

The area of tea cultivation is given in *mu*. One *mu* is equivalent to approximately 0.0667 hectares.

Puer Tea

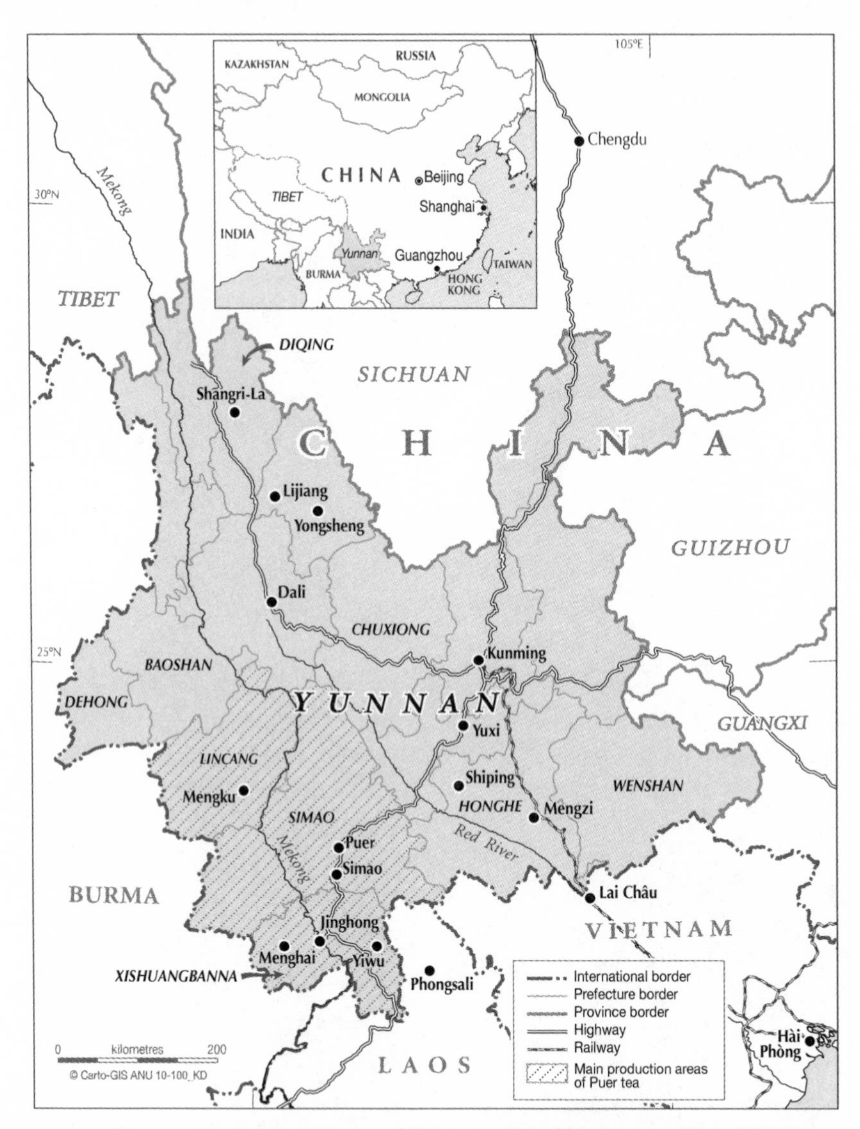

MAP I.1 Yunnan. Area: 394,000 square kilometers. Population: 44.5 million (CADN 2010a). The Han make up almost 70 percent of Yunnan's population. Apart from the Han, there are twenty-five ethnic groups (according to the government's classification), the largest number in any province in China.

MAP I.2 Xishuangbanna Dai Autonomous Prefecture in Yunnan. Area: 20,000 square kilometers. Population: 840,000. Apart from the Han, there are thirteen ethnic groups, including Dai, Hani, Lahu, Bulang, and Jinuo (according to the government's classification), who together make up almost 74 percent of the population. Xishuangbanna is divided into three administrative parts: Menghai County, Jinghong City, and Mengla County (CADN 2010b).

Introduction

Tea has been familiar to me since my childhood in Kunming, Yunnan, in southwest China. At my family's house there was always a tea jar. We drank tea often, though not necessarily every day or with every meal. Tea more than two years old or so had to be thrown away. As a child, I was taught to make tea whenever a guest came to visit. I would simply put some loose tea leaves in a glass and pour in hot water. The resulting brew would be a yellow-green color, and everybody recognized this as Yunnan's green tea. Once the guest had drunk some of the tea in the glass, my parents would tell me to add more water so that the brewed tea would not become too strong, as the tea leaves were still infusing. The guest might not be thirsty, but he would take frequent sips nonetheless. My parents and the guest would talk about something apart from tea. In these conversations, tea was both important and unimportant.

Like many people at that time, my parents and I were unconcerned about the difference between Yunnan's Puer tea and green tea. My understanding was that green tea was the loose tea stored in a jar that we served to guests, while Puer tea (also spelled Pu'er or Pu-erh, and pronounced in two syllables) was compressed, usually into a bowl shape (*tuo cha*). The latter was more often given as a gift to friends outside Yunnan than consumed at home. I once found some leftover Puer tea in a cabinet. Each cake was shaped like a small bowl, half the size of a ping-pong ball. Out of curiosity, I took a cake and infused it. The color of the brew was similar to that of Yunnan's green tea. But the compressed small bowl unexpectedly swelled up in the hot water to more than five times its original size. The brew was so strong that I decided I didn't like it.

I didn't try Puer tea again until 2002, when I joined a film crew that was making a documentary about people involved in tea production in Simao and Xishuangbanna, two of the tea production areas in Yunnan (map I.1). The film director from Beijing continuously drank Puer tea that he had bought in Yunnan. The dry compressed tea and the brew he made were

FIG. I.1 Two kinds of loose raw Puer tea (*maocha*). Photo by the author.

FIG. I.2 Various shapes of compressed Puer tea sold in the market. Photo by the author.

both dark red. I asked for a taste. It was quite smooth, but it had an earthy smell. Another member of the crew, who was also from Yunnan, said it was moldy. Its color, smell, and taste were all new to me. The director said he drank the tea to control his high blood pressure. We later visited Yiwu, a township in Xishuangbanna known for its Puer tea (maps I.1–2). There, local people showed us how to produce compressed cakes of the tea. This tea, in contrast to that which the director drank, had a "sunny" smell, and its yellow-green color resembled that of Yunnan's green tea.

Back in Kunming, the capital city, I soon heard more about Puer tea. Compressed into the form of a cake (*bing cha* or *yuan cha*), bowl (*tuo cha*), or brick (*zhuan cha*), it was selling exceptionally well at the time (figs. I.1–2). I learned that there were two kinds of Puer tea: the green kind was raw tea (*sheng cha*) and the dark kind was artificially fermented tea (*shu cha*) (figs. I.3–4). In addition, I was shown a third kind that was said to have been aged for over five years, sometimes for as long as several decades. This aged tea was developed from the first two kinds and was much more expensive. A popular saying was applied to Puer tea: "The longer it's stored, the better it tastes" (*Yue chen yue xiang*). In other words, "The older the better." The most precious Puer tea, I was told, was that from Yunnan, which is now collected by connoisseurs in Hong Kong and Taiwan. Puer tea made from older tea plants was also valued more highly, although this information was less accessible to ordinary consumers. Connoisseurs differentiate between forest tea (*da shu cha*) and terrace tea (*tai di cha* or *xiao shu cha*).[1] Forest tea is tea produced from tall tea trees—often over one hundred years old—sheltered by the forest canopy, initially cultivated by ethnic minorities such as the

FIG. I.3 *(above)* A brew of raw Puer tea. The one-year-old tea was stored in Kunming. Photo by the author.

FIG. I.4 *(right)* A brew of artificially fermented Puer tea displayed at the 2007 Kunming tea trade fair. The color of very aged Puer tea is also dark red, but the taste is different. Photo by the author.

Bulang and Hani in Yunnan. Terrace tea refers to narrowly and regularly arranged tea bushes, a style of cultivation that had begun in Yunnan only in the late 1970s and the early 1980s.[2] In 2007, the price of forest Puer tea was four or five times that of terrace tea, because the former was considered to come from a more ecologically healthy environment and was thought to taste much better.

Whatever its specific kind, Puer tea—as profiled in popular books and in the market—was both attractive and perplexing. It was said to have had greater medicinal functions and cultural values than any other type of tea. It also had multiple definitions. It was repeatedly authenticated for ordinary consumers, who, nevertheless, were often left confused. The most serious problem was that there were too many counterfeits: Puer tea that had been aged for three years was said to have been aged for thirty; tea material originating in Sichuan was said to have come from Yunnan; terrace tea was labeled as forest tea; and the distinction between green tea and raw Puer tea was difficult to discern. The information on a package of Puer tea was often unclear, and interpreting it correctly depended upon the consumer's knowledge and careful negotiations with traders.

Even though the criteria for authenticating Puer tea remained obscure, it was continually celebrated. A growing number of "tea experts" were writing about and trading Puer tea. The mass media took advantage of its popularity, and the government of Yunnan declared it a provincial symbol, supporting all sorts of tea propaganda. In Kunming teahouses, Puer tea was infused in delicate tea sets. Tea-tasting events, which had previously been seen only in ancient paintings and literature, flourished. People met over Puer tea, talked about Puer tea, and competed to be the most knowledgeable about Puer tea or to produce the highest-quality Puer tea. In order to find out the "truth" about Puer tea, urban dwellers launched journeys into rural production areas. These activities boosted the rural tea economy, changing the system of agricultural production deeply.

As a result of all these efforts, the profile and price of Puer tea peaked in early 2007, only to drop unexpectedly in May of the same year. Before and after this, many people felt pleased or worried, gaining or losing, struggling or relaxing, all because of Puer tea.

How was Puer tea transformed from something ordinary and familiar into something remarkable and exotic in Yunnan? Why have so many people been compelled to collect, drink, admire, and study it? How did multiple players cause the value of Puer tea to skyrocket and then plummet? Why have counterfeits flourished despite so many appeals for regulation? And how do ordinary tea peasants, traders, and consumers survive in this chaotic battle? This book reflects upon how Puer tea has been packaged by multiple actors into a fashionable drink with multiple authenticities, and, more importantly, how this packaging has been challenged and unpacked by multiple counterforces. Through looking at the packaging and unpackaging process, it explores the temporal interaction between China's Reform period (formally since the early 1980s) and the past; the spatial interaction between Yunnan and other tea production and consumption areas in China and overseas; and the social interaction between various Puer tea actors. These interactions are taking place at a particular moment when China is accelerating its economic and cultural development. New production and consumption trends have appeared. But rather than looking at them as completely new, this book suggests that some of them have involved transformation and repackaging of long-standing elements of Chinese culture. The so-called "antique fashion" for Puer tea mirrors the enduring characteristics of Chinese cultural consumption, though such characteristics are often presented with new forms and new meanings. In particular, this book explores the Chinese concept of *jianghu* (lit. "rivers and lakes," a nonmainstream space popularly used for swordsman culture) to interpret the flourishing of counterfeits and the unavailable authenticity of consumer goods in China, as exemplified by the case of Puer tea.

This book traces Puer tea's "cultural biography" (Appadurai 1986)—an examination of the politics linking a commodity's value and exchange to its detailed social biography—from Yunnan out, from multiple rural production sites to multiple urban consumption areas. One of my main research sites was Yiwu township in Xishuangbanna, Yunnan, whose history of producing Puer tea exemplifies the packaging and unpackaging process of Puer tea throughout Yunnan. Sometimes the narrative shifts to Menghai and Simao, two other important production places in Yunnan, to describe special events, particularly when these multiple production sites competed over authenticity. My other main research site was Kunming, the locus of Puer tea's distribution and consumption in southwest China. I also made short visits to Hong Kong, Guangzhou, and Taiwan, where the consumption of Puer tea has had a major impact on Yunnan. While following the trajectory of Puer tea, I also followed tea traders. Therefore, some actors who appear in rural tea fields collaborating and competing with the locals also appear in urban teahouses, where they drink tea and engage in debates about authenticity.

Following these actors' journeys from the rural to the urban, and following Puer tea's rise and fall in 2007, chapters feature a seasonal theme inspired by other writings on tea (Huang Anxi 2004) and Chinese consumption culture (Brook 1998). The framework emerges from the belief put forward by Chinese traditional literature, medical science, and Daoism (Taoism) that one needs to "nourish life" (*yangsheng*) in accord with the seasons: sprouting in spring (*chun sheng*), growing up in summer (*xia zhang*), harvesting as well as withdrawing in autumn (*qiu shou*), and storing and hiding in winter (*dong cang*). For example, a prominent theme in Chinese literature is "lament for autumn" (*bei qiu*), as voiced by the poet Song Yu in the third century B.C.E.: "Alas for the breath of autumn! Wan and drear! Flowers and leaf fluttering fall and turn to decay" (Hawkes 1957: 92). In the "Autumn" part of this book, this theme coincides with local worries about the recession in the Puer tea market.

A set of short documentary films illustrates the landscapes of tea processing and people's sensory experiences in tea tasting (see appendix 2). Rather than simply paralleling the text's narratives, the films provide their own clues and themes. For instance, several films are based on the lives of family tea producers and traders who are not examined in detail in this book. The films thus provide alternative narratives, as do the voices and alternative resolutions adopted by Puer tea actors.

ALL PUER TEA IS FROM YUNNAN

The Chinese were the first people in the world to domesticate tea-producing *Camellia sinensis* plants and begin drinking tea (Zhu Zizhen 1996; Mair and Hoh 2009). According to folklore, tea was initially discovered by Shen Nong (lit., "Divine Farmer"), a legendary Chinese tribal head and pioneer of Chinese agriculture and medicine. Shen Nong was once poisoned by toxic herbs but saved himself by drinking a brew made from tea leaves (Zhu Zizhen 1996; Lu Yu 2003). Textual records indicate that a specialized tea market had arisen in Sichuan as early as the sixth century B.C.E. By the third century C.E., tea had become a popular drink in what is now southern China, and during the Tang dynasty (618–907), tea drinking became prevalent all over the empire. The first known book about tea, *The Classic of Tea* (Cha jing) by Lu Yu (2003), was produced in the eighth century (Goodwin 1993; Zhu Zizhen 1996; Guan Jianping 2001). Compressed tea was popular initially, but loose tea began to dominate after the fourteenth century, when it is said that the first emperor of the Ming dynasty (1368–1644), Zhu Yuanzhang (r. 1368–1398), outlawed compressed tea out of concern for the burden it placed on tea producers.[3] The emperor's order, however, didn't reach remote Yunnan, which, at that time, was not under the complete control of the imperial court. As a result, Yunnan's Puer tea continued to be produced in compressed form, which is also more convenient for long-distance trade. Most recent popular writings on Puer tea, many of them composed by Yunnanese, stress that Chinese textual records on tea have long omitted Yunnan, a frontier region far away from central China, despite its significant role in tea production and consumption (see Lei Pingyang 2000; Zhou Hongjie 2004).

Many scholars agree that Yunnan (map I.1), in southwest China, is one of the most important places in the history of tea.[4] Along the Mekong River in southwest Yunnan, there are plentiful and excellent tea tree resources, mainly located in three subdistricts: Xishuangbanna, Simao, and Lincang.[5] Ethnic minority groups such as the Bulang, Deang, Wa, Hani, and Jinuo are thought to have cultivated tea for at least one thousand years (Zhang Shungao and Su Fanghua 2007; Li Quanmin 2008).[6] Han immigrants didn't come to Yunnan in large numbers until the fourteenth century, and it was not until the early eighteenth century that Chinese merchants began entering the tea-growing areas (Giersch 2006: 24–25). The Chinese came to dominate the tea trade between Yunnan and inland China and the neighboring Southeast Asian regions in the late nineteenth and early twentieth centuries (Hill 1989). And as many Chinese scholars stress, Han immigrants also introduced new tea production techniques and consumption customs into Yunnan (see Lin Chaomin 2006).

Most tea trees in Yunnan are *Camellia sinensis assamica.* This type of tea is referred to as large-leaf tea, in contrast to *Camellia sinensis sinensis,* or small-leaf tea.[7] Current tea science suggests that tea plants in other parts of China—most of which are of the small-leaf variety—evolved from the large-leaf variety prevalent in Yunnan (Chen Xingtan 1994). The large-leaf variety is acknowledged to be the most suitable for making Puer tea. But Puer tea does not refer to a particular category or species of tea tree. Instead, it was named after the town called Puer, a center for goods distribution and taxation in southern Yunnan since the early seventeenth century (Fang Guoyu 2001: 427–428; Xie Zhaozhi 2005: 3; Ma Jianxiong 2007: 563).

Like Puer tea, many kinds of Chinese tea were initially named after places, whereas contemporary tea science names and categorizes tea mainly on the basis of production techniques. There are six kinds of Chinese tea, distinguished by processing methods, especially fermentation (Chen Chuan 1984, [1979] 1999). Tea fermentation usually refers to the oxidation reaction (from contact with the air) or microbial enzymatic reaction (from stacking in a moist environment).[8] One important procedure in tea processing is the activation or suppression of fermentation, and different degrees of fermentation produce different flavors (Cai Rongzhang 2006). The different categories reflect the color of the resulting tea brew: green tea, yellow tea, white tea, blue-green tea, red tea, and dark tea. Within each category are numerous subcategories, determined by trivial differences in the tea plant and processing. Green tea (*lü cha*), usually including jasmine tea (scented green tea), is nonfermented.[9] Depending on the different techniques of fermentation suppression and the drying process, green tea can be subdivided into sun-dried green tea (*shai qing*), steamed green tea (*zheng qing*), stir-roasted green tea (*chao qing*),[10] and baked green tea (*hong qing*). Yellow tea (*huang cha*) is also nonfermented, and it is only slightly different from green tea. White tea (*bai cha*) is very lightly fermented. Blue-green tea (*qing cha*), such as oolong, is semifermented. Red tea (*hong cha*), which in English is referred to as black tea, is fully fermented. Finally, dark tea (*hei cha*) is also fully fermented, but its fermentation happens later than that of red tea. According to this six-fold system of classification, Puer tea is categorized as dark tea (Chen Chuan 1984). But when Puer tea became popular at the beginning of the twenty-first century, its proponents called for it to be treated as an independent category (table. I.1).

As Puer tea's definition is debated, it is hard to tell what exactly Puer tea is, although some aspects are less controversial than others. Contrasting views are presented as necessary here.

At present, Puer tea can be categorized into three types depending on their postfermentation. The first category is raw Puer tea (*sheng cha* or *sheng pu*), which is made using large-leaf tea leaves and can be very astringent when young. Some tea experts argue that postfermentation is a key characteristic in Puer tea's processing, but raw Puer tea hasn't undergone any postfermentation at all and is very similar to green tea (Zou Jiaju 2005). Raw Puer tea can be compressed into cake, bowl, brick, or melon form, or it can be left in its loose form, which is often called *maocha*.

The second category is aged raw Puer tea (*lao sheng cha*), which should be at least five years old, although clear agreement hasn't yet been reached on how many years' storage is required before the tea is considered aged. Generally, the older the tea, the higher its price. The staggering prices of aged Puer teas in Hong Kong and Taiwan—such as a seventy-year-old cake (357 grams) that sold for more than ¥1 million—has inspired the production and storage of more Puer tea. It is believed that "natural" fermentation (mostly oxidation, possibly also with some microbial enzymatic reaction) occurs during long-term storage, turning the tea from astringent to mild. But "natural" is a relative concept, because some people also create a humid storage environment to accelerate fermentation, which brings this second category close to the third category.

The third category is artificially fermented or "ripe" Puer tea (*shu cha* or *shu pu*), which is the product of a different method used to mellow the astringency of raw Puer tea. By subjecting *maocha* to a specific temperature and humidity level, postfermentation (mainly microbial enzymatic reaction)

TABLE I.1 Categories and production process of Chinese tea

Fermentation Type	General Tea Category	Subcategory or Examples		Production Process
Nonfermented	Green tea	Kill the green by steaming	Japanese green tea; Yunnan Steamed Enzyme tea (Zhengmei)	Harvest—kill the green—roll—dry (—scented: jasmine tea; osmanthus tea)
		Kill the green by stir-roasting	Dragon Well; Raw Puer tea (*sheng cha*)	
	Yellow tea	Jun Mountain Silver Needle (Junshan Yinzhen); Meng Mountain Yellow Buds (Mengding Huangya)		Based on green tea—sealed yellowing (*men huang*)
Partially fermented	White tea	Silver Needle White Hair (Yinzhen Baihao); White Peony (Bai Mudan)		Harvest—wither—dry
	Blue-green tea	Iron Goddess of Mercy (Tieguanyin); Wuyi Rock tea (Wuyi Yancha); White Hair Oolong (Baihao Wulong)		Harvest—wither—ferment—kill the green—dry
Fully fermented	Red tea	Keemun (Qi Hong) Red tea; Yunnan Red tea (Dian Hong); Earl Grey		Harvest—wither—roll—ferment—dry
	Dark tea	Anhua Dark tea; Liubao tea		Harvest—wither—kill the green—roll—ferment—dry
	Puer tea (fermented)	Artificially fermented Puer tea (*shu cha*)		Harvest—wither—kill the green—roll—dry—ferment—dry
		Naturally fermented Puer tea / aged raw Puer tea (*lao sheng cha*)		Based on raw Puer tea—age

Note: This categorization and description is based upon Chen Chuan (1984); Zou Jiaju (2004); Cai Rongzhang (2006); Mair and Hoh (2009). Also see table A.2 for more details about the Puer tea production process.

can be completed within two or three months. This technique was formally invented in Kunming in 1973. In the tea market, it is said that artificially fermented Puer tea can also be further "naturally" stored for a long time, resulting in what would be called aged artificially fermented Puer tea.

Usually tea—especially green tea, the dominant variety consumed by the Chinese—is appreciated for its freshness. Puer tea, by contrast, has come to be valued in the late twentieth and early twenty-first century for the flavor it develops as it ages. The origin of aging Puer tea is in contention, though it probably arose during long-distance trade and the interaction between Yunnan, the production region, and the various places where the tea is consumed.

Before the emergence of modern transportation, Puer tea was carried by horse or mule caravan from Yunnan, a rugged mountainous region, to the outside world. This transport was dominated by Yunnanese Chinese (both Han and Muslim Chinese), but Tibetans also joined in some parts of the journey (Hill 1989). Puer tea had become famous in Tibet, Beijing, and Hong Kong, where people drank it to help them digest greasy food. What was carried initially was very young raw Puer tea, often made into a compressed form. Legend has it that the benefits of postfermenting Puer tea were accidentally discovered as a result of this caravan transport; the sunshine and rainfall affected the tea's flavor and tamed its astringency (Su Fanghua 2002: 50; Mu Jihong 2004: 92; Zhou Hongjie 2004: 8). Another group of tea commentators, mainly from the Pearl River Delta,[11] however, argue that the Cantonese first discovered that Puer tea's flavor improved and turned mild after being stored for several years (He Jingcheng 2002: 118–125).[12] It's impossible to know which story is accurate, but it is true that aged Puer tea is rarely found in present-day Yunnan and that many Yunnanese traders buy the tea from Hong Kong and Taiwan.

Although who discovered the benefits of aging Puer tea remains unresolved, the Yunnanese are taking strategic action to associate Puer tea with Yunnan. In a sense, this echoes the recent academic examination of the formation of "Yunnan" and its unique status in a global context. Historically, several indigenous regimes existed in the area that now comprises Yunnan. These included the Dian Kingdom in the third century B.C.E., the powerful Nanzhao Kingdom in the mid-seventh century, and the Dali Kingdom in the early tenth century. Although the Yuan Mongols' conquest in 1253 initiated Yunnan's incorporation into the Chinese empire, for a long time Yunnan was not fully incorporated but was made up of multiple independent or semi-independent local regimes that had closer relationships with neighboring Southeast Asian areas; nor were all the present parts of Yunnan called "Yunnan" (Giersch 2006; Hill 1998; Yang Bin 2006).

Although eliding some of these "sensitive" political and ethnic issues, Yunnanese writers ask their audience to pay close attention to Yunnan's history in order to understand Puer tea. They highlight Yunnan's essential role in the development of Puer tea by borrowing insights from academic

FIG. I.5 Lahu people under their tea trees in Menghai, Xishuangbanna. They are wearing traditional costumes for the local government's tea event in March 2007. Photo courtesy of Sun Jingfeng.

research, stressing that Yunnan has never been isolated by high mountains or tough roads. They argue instead that it has long been closer to Southeast Asia in terms of geography, economics, ethnicity, and religion, and that it has played an important role in communication and cultural exchange between India, Southeast Asia, inland China, and the wider world (Mu Jihong 1992; Lei Pingyang 2000; Mu Jihong 2004; Ruan Dianrong 2005a; Zou Jiaju 2005). Puer tea is portrayed by these writers as a key product that embodied Yunnan's relative autonomy and its contribution to the world. The images of multiple ethnic groups in Yunnan, with their special costumes and customs, and the stories about past tea caravans become unique symbols representing Puer tea's value and the unique status of Yunnan (fig. I.5–6). This shows a mystifying and complicating trend, in which Yunnan is described as a pure and mysterious place and Puer tea is endowed with multiple values representing Yunnan's rich culture.

Nevertheless, these writers also simplify history and reality, consistently with current administrative organization and government policy. In this regard, their voices become Sinocentric. They consider all the existing land in the province as incontrovertible parts of Yunnan, and hence of China.

FIG. I.6 Tea caravans are still in use in some parts of Yunnan. Photo by the author.

While they stress that Yunnan contains multiple ethnic groups and cultures, they also make it clear that Yunnan is a "united" province of China, ignoring its complex history.

TRANSFORMATION AND THE DESIRE TO BALANCE

How has Puer tea been endowed with so many symbolic meanings? In what social context was its recent popularity generated? These symbolic meanings are strategically inherited from those applied to Chinese tea generally over a long period of time; more importantly, the construction and application of these symbolic meanings is taking place at a critical moment in China's transformation in politics, economics, and consumer culture. The Puer tea fad symbolically represents multiple desires that contrast the past with the present, as well as different places and actors.

Anthropologists and sociologists investigate attitudes toward food from diverse perspectives. One of the most important approaches links symbolic meanings of food to issues of identity. Beginning with the work of Claude Lévi-Strauss (1970), scholars have acknowledged that food is not only good to eat but also good to think with. Lévi-Strauss developed the binary distinction between the raw and the cooked, stressing that artificial interference is the key factor that transforms food from the natural to the cultural. As we will see, the case of Puer tea sometimes departs from this general scheme.

The symbolic meanings of food have been explored in many different contexts. Some scholars have considered the role of food in spiritual contexts, as certain food can become an obvious marker of a certain religion (Toomey 1994; Feeley-Harnik 1995). Some stress that food plays an important role in memory because it can awaken one's senses to remember home or the past (Seremetakis 1994; Sutton 2001). Some attribute different styles of food consumption to divergent tastes and class positions (Goody 1982; Bourdieu 1984). Some look at how food has been taken as one inevitable part of tourism to signify different travel "flavours" (Heldke 2005; Germann-Molz 2004). Some discuss food customs from the point of view of gender, as throughout history certain modes of food consumption have been considered masculine or feminine (Counihan and Kaplan 1998). Some have also explored how food is tied to ethnicity and used to represent distinct ethnic identities, even with spatial and temporal change (Ohnuki-Tierney 1993; Tam 2002). Chinese food is a particularly rich subject, as it allows us to explore how unique Chinese concepts are embodied in food and how certain imported foods are locally reinterpreted (Anderson 1980, 1988; Watson 1997; Wu and Cheung 2002; Su Heng-an 2004; Sterckx 2005).

Emiko Ohnuki-Tierney (1993) writes about the Japanese use of rice as a "metaphor of [the] self," arguing that the symbolic meanings developed for certain staple foods are the sediments of "historical process." Over a long period of time these meanings become "natural" to the people of the nation (1993: 6). Although rice has become the dominant representation for Japan, it is not necessarily quantitatively important to all Japanese, and in fact there are large portions of the Japanese population who don't depend on rice to survive. Nor is the identification of the Japanese with rice a conscious presentation. The customs and metaphors relevant to rice are merely applied by the general population in their daily lives. This is different from the consciously symbolized and abstracted meanings developed by cultural interpreters such as anthropologists (Ohnuki-Tierney 1993: 5).

In many respects, the case of tea in China is similar. Its symbolic meanings have developed over a long period and have become naturalized as a part of China's history. Tea has become the exclusive national drink, though it's not actually consumed by the entire population. Moreover, it is regarded as far more than a drink to quench thirst. In ordinary people's lives, tea is served to guests for hospitality and is essential in managing social relations. It is commonly given as a gift to respected people, especially during festivals. In the past, many Chinese people used tea as a betrothal gift during wedding negotiations. Tea is also used to negotiate between the mundane and the sacred—for instance, as an important offering during ancestor worship rituals. Indigenous ethnic groups in Yunnan, such as the Hani, actually worship tea trees and don't allow them to be destroyed at random (Shi Junchao 1999; Xu Jianchu 2007).

However, whereas the symbolic meanings of rice appear to have been mainly unconsciously developed in Japanese daily lives, the Chinese literati have played a role in consciously highlighting the cultural importance of tea, like the French gentry's role in boosting the status of French cuisine (Ulin 1996; Ferguson 1998). Tea drinking is linked by the literati to other forms of art—such as poetry, calligraphy, and painting—and it represents a frugal, pure, and elegant lifestyle (Shen Dongmei 2007). Many ancient landscape paintings portray human figures in nature drinking tea, indicating escape from political disputes and enjoyment of freedom in a self-constructed utopia (fig. I.7).

Tea has also been linked by the literati to important forms of Chinese religion. Essays and poems often associate tea with frugality, Confucian benevolence, and moderation. In the fourth century, tea banquets began to be advocated as alternatives to extravagant alcohol banquets, and tea was taken as a sobering drink in contrast to alcohol, which often led to disorder (Guan Jianping 2001). Chinese tea was also promoted for its medicinal functions: refreshing the mind, aiding digestion, reducing fever, and promoting urination. These functions were further promoted by the Daoist idea of "nourishing life" (*yangsheng*), or improving health through proper eating and thinking. Tea also became important in temples, where Buddhist monks drank it to refresh their mind and assist in meditation, as expressed in the saying "Tea and Ch'an Buddhism have the same flavor" (*cha chan yi wei*) (Benn 2005).

It is not always easy to differentiate between the metaphoric meanings

FIG. I.7 Zhao Yuan (14th century). *Lu Yu Brewing Tea* (Lu Yu peng cha tu). Ink and watercolor on paper, 27 × 78cm. National Palace Museum, Taipei.

of tea used by folk audiences and those used by the cultural elites. Quite often the latter develops its further insights from the former, and then these insights flow back into popular culture and commingle. This explains why many ordinary Chinese also believe that tea drinking brings about good order and enjoyment. In other words, the symbolic meanings of tea in China are actually generated by both conscious and unconscious representations, and naturalized by the strength of contributions from both folk and elite culture.

The uneven segments in this historical process—namely, the variation, contrast, transformation, and subversion in the process of symbolic identification and the counterforces that challenge established symbolic meanings—deserve particular attention. Chinese symbolic identification with tea has not always naturally or steadily developed throughout history; rather, it has been uneven and shaped by political and economic pressures and social demands.[13] Before and after the Reform era, the profile of tea in China was quite different. During the time of Mao (1949–1976), and especially during the Cultural Revolution (1966–1976), it was suppressed by the exclusive emphasis on political struggle. In this period, living standards were low (fig. I.8) and consumption-based lifestyles were condemned as "negative capitalism" (*ziben zhuyi weiba*) (see also Zeng Zhixian 2001: 93). The positive meanings of tea were buried, the number of public teahouses was greatly reduced, and tea consumption was largely limited to the family and work unit. The Reform era, which formally started in the early 1980s, saw a gradual elevation of living standards. Various forms of entertainment were encouraged, especially after the mid-1990s, when China began its economic surge. Enter-

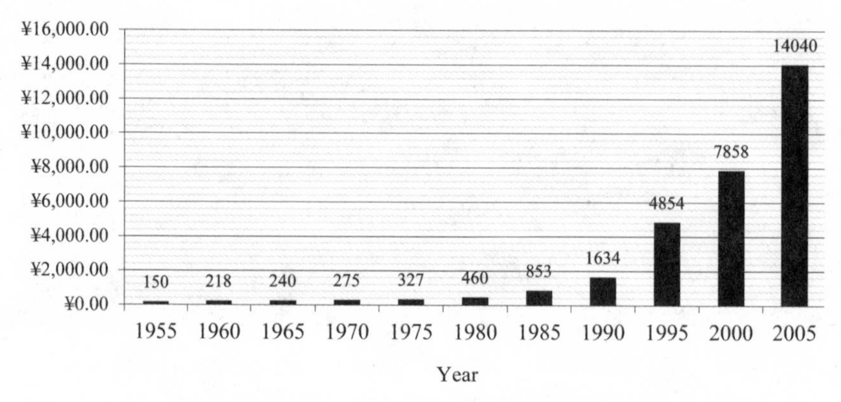

FIG. 1.8 China's GDP per capita (1955–2005). Source: NBSC (2009).

ing the twenty-first century, the so-called "consumption revolution" has become more intensely staged (Davis 2000; Latham, Thompson, and Klein 2006). Tea is once again stressed as an essential national representation. All sorts of tea events—such as tea auctions, tea-serving performances, and tea-tasting competitions—have sprung up, and these are often elevated to the status of art. These events reflect an extraordinary craze for tea culture in contemporary China (D'Abbs 2009; Tan and Ding 2010). One survey reveals that by the end of 2006, there were a total of four thousand wholesale, retail, and service tea establishments in Kunming, in contrast to the low profile of public tea services during the Mao era.[14] It is amid this transformation that Yunnan's Puer tea has become popular. Remarkably, all the current teahouses in Kunming sell mainly Puer tea, a tea that was consumed by Yunnanese in the past, but only recently appreciated for its aged value.

The present cultural packaging of Puer tea has inherited and borrowed many elements from the packaging of Chinese tea more generally. For example, popular writers on Puer tea argue that Puer is the only kind of tea that maintains "the legacy of Tang and Song," as it is still processed into a compressed form (Lei Pingyang 2000: 1–3; Deng Shihai 2004: 24).[15] The other distinctive feature of Puer tea is its link to time. Many writers say that a piece of Puer tea won't be good until it has been aged for a long time by means of a natural process, whether along the caravan route or in storage, just as a man won't be mature and wise until he has had enough life experience (Li Yan and Yang Zejun 2004; Ruan Dianrong 2005b). Hence, some writers and drinkers declare that superior Puer tea is naturally aged, which echoes the

Daoist appreciation of spontaneity and "flavorless flavor" (*wu wei zhi wei*) (Deng Shihai 2004: 49).[16]

Whether related to individual identity or Yunnanese group identity, these symbolic meanings of Puer tea are consciously constructed. On the one hand, many Yunnanese are confused by them, especially by the sudden appreciation for the flavor of aged Puer tea, which ironically was "artificially" created by a group of advocates in only about five years, in contrast to the slow aging process. Other studies have also documented the rapid invention of foodways in China (Kyllo 2007) and elsewhere (Haverluk 2002; Hsü Ching-wen 2005), while the uniqueness of Puer tea's case lies in the depth of its packaged values and the widespread counterforces that unpack and deconstruct the newly established values. Moreover, examining the unprecedented packaging of Puer tea would contribute to the study of how values and meanings are created through transnational or transregional commodity networks (see Friedberg 2004; Wilk 2006; Belasco and Horowitz 2009), an underexplored topic in relation to Chinese food and globalization (though see Tagliacozzo and Chang 2011).

On the other hand, these consciously constructed meanings for Puer tea reveal a certain kind of "historical process." When put on a historical timeline, the contrast between these newly constructed meanings and past meanings indicates that meanings attached to Puer tea have developed at specific points in China's historical development. Just as the Japanese use of rice as a metaphor for self is, in Ohnuki-Tierney's words, "born through discourse with the other" (1993: 8), the packaging of Puer tea's symbolic value happened in the context of Reform China's transformation. New meanings are constructed through discourse and contrast with the historical past.

In relation to this temporal discourse, consumers, situated in a context of transformation and adaptation, desire to consume something both new and old in order to counterbalance the unforgettable past. First, they want to become wealthy, to overcome the poverty of the past. Those who have become rich are eager to find channels for investment, and Puer tea, the "drinkable antique" valued for its aged taste, is a good candidate. As one slogan says, "You will regret tomorrow what you don't store today."

Second, consumers want to live healthy lives, to achieve physical balance for their bodies, which had been neglected during years of poverty. This echoes the consumption trend globally (Hollander 2003) and reflects distinctive Chinese beliefs that food is also medicine and that eating is essential to balance yin and yang, or cold and hot, in the body (Anderson 1980; Belasco and Scranton 2002; Ismail 2002; Su Heng-an 2004). The advertising claim that Puer tea is more effective than any other tea for helping with digestion is widely accepted, and some consumers seek out Puer tea gathered from forest trees due to its superior ecological, taste, and health value. In both urban teahouses and rural production areas, people drink Puer tea before meals to stimulate the appetite or after meals to counteract greasiness. It is hard to say whether they drink Puer tea in order to eat more or eat more in order to drink Puer tea. In other words, the Puer tea boom is part of the ongoing embrace of fine Chinese cuisine and the pursuit of good health.

Third, consumers want to revive the past (*fugu*) to compensate for the suppression of interest in collecting antiques during the Maoist era. Spurred by economic growth, the expanding upper and middle classes have more income and leisure time to seek out and appreciate traditional Chinese aesthetic values, such as antique paintings, calligraphy, and other artistic objects. They also seek to balance the ongoing modernity and globalization that make many old things disappear. Although an antique, Puer tea is said to be "still alive," and it can be consumed even after a long period (Deng Shihai 2004: 34). This caters to the need of the present generation to obtain something both old and new, in a "commoditization of nostalgia" (Sutton 2001: 163)—a nostalgia that has been created as a result of twentieth-century tourism (Lowenthal 1985). Moreover, it also reflects the so-called "culture fever" (*wenhua re*) phenomenon that has swept China since the mid-1980s (Wang Jing 1996; Schein 2000).

Fourth, consumers desire authenticity: authentic Puer tea, authentic identity, and authentic lifestyle. They wish to achieve distinct identification to balance the Maoist era's emphasis on unification. In a sense this quest for authenticity reflects the trend for a new kind of individualism. It is also a response to the uncertainty brought about by modernity. Consumers believe that, in order to find authentic Puer tea, they must go to the rural tea mountains, which are unpolluted, quiet, and slow paced, in contrast to the polluted, noisy, and fast-paced urban life. Taking a cue from the "slow food movement" in Europe (Leitch 2003), aged Puer tea is seen as a "slow beverage," used to counterbalance the rapid pace of modernity (Ruan Dianrong 2005b; Zhu Xiaohua 2007). In this regard, authentic Puer tea is identified with an authentic lifestyle, and discovering authentic Puer tea represents one's ability to achieve the lifestyle that one really wants; authenticity is endowed with meanings of self-determination and freedom, and it represents the desire to find certainty in the midst of uncertainty. In turn, the changeable and debatable authenticity of Puer tea reflects the ambiguity about the sort of lifestyle that is most desirable. As anthropologists have argued, the ways people construct biographies of things reflects the way that they construct their own identities; in other words, they construct their own identities with the constructed biographies of things, and the complex and conflicting biographies of things reflects the uncertainty of people's identities (Kopytoff 1986).

In a period of transformation, the symbolic meanings of Puer tea are thus being packaged in order to counterbalance the impoverished past. Symbols associated with Puer tea represent new national, regional, and individual identities to counterbalance prior identities and ongoing globalization. Countervoices also emerge to further counterbalance these newly constructed "authentic" meanings. The packaging and unpackaging has happened quickly, and the contrast between these alternative voices raises questions as to how long some symbolic meanings of Puer tea can last, leading to a series of paradoxical feelings about the production and consumption of Puer tea.

THE *JIANGHU* OF PUER TEA AND HANDCRAFTED AUTHENTICITY

Despite people's desire for authenticity, fake Puer tea flourishes. A popular saying tells consumers that 90 percent of Puer tea in the market can't be authentic. Although the provincial government has established a series of regulations, the existence of so-called fake Puer tea, like other fake products in China, continues. While some people complain that there is no clear and strong standard for authenticity, others are unconcerned because they can survive without such a standard. Many tea experts compare the world of Puer tea to that of a *jianghu,* referring to the chaotic situation of Puer tea. This kind of casual comment was formally taken up by the magazine *Puer Jianghu,* which was established in April 2007. The editor-in-chief told me that since there had been so many disputes concerning Puer tea, he would like to see the magazine focus more on fun for a change (fig. I.9). The meanings of *jianghu,* historically and contextually, can explain how the tapestry

FIG. I.9 A page in the opening issue of *Puer Jianghu* (2007b). The original image is Lin Chong, the coach of 800,000 imperial guards in *The Water Margin*. Here his title is changed to "coach of the Puer *Jianghu*." The caption on the left says, "Puer is a *Jianghu*. No moon, no stars, but only full of the shadow of swords. *Jianghu* is not awful. What's awful is uncertainty and wandering alone." Photo by the author.

of Puer tea is woven by multiple actors and the authenticity of Puer tea is contextually packaged and counterpackaged.

Jianghu literally means "rivers and lakes." Its deeper meaning encompasses nongovernmental space, in a sense echoing James Scott's (2009) description of Yunnan as part of "Zomia," a mountainous region where people have sought refuge from the power of the state. The early use of *jianghu* contained meanings about reclusion, as in the remark by Han historian Sima Qian (145 B.C.E.–87 B.C.E.), who said, "[After fulfilling his task, Fan Li resigned and] took a small boat, floating in the *jianghu*" (Sima Qian 2011: 2459). In this regard, *jianghu* is inhabited by hermits who share the same interests as those characters portrayed in landscape paintings drinking tea, both going beyond the political arena and enjoying what John Christopher Hamm describes as "individual liberty given solace and substance by romantic fulfilment on the one hand and transmitted cultural practices on the other" (2005: 137).

More popularly, the world of *jianghu* appears in Chinese martial arts fiction. Since the Tang dynasty (seventh to ninth century), when martial arts fiction flourished,[17] *jianghu* has been depicted as a utopia that partially reflects reality, in which Chinese knights-errant or wandering swordsmen (*xiake*) travel around, compete in martial arts (*kungfu*), and go beyond the political control of the court to help the poor and the oppressed, though not declaring their direct defiance of the court (Liu 1967; Chen Pingyuan 1997; Jing Wendong 2003). While evading the court's complexity, this *jianghu* of knights-errant has its own chaos, full of dangers and contests, with battles occurring at inns, waterways, mountains, temples, and deserts. This theme is what most contemporary people refer to when they mention "the *jianghu* of Puer tea."

In some cases, *jianghu* becomes a space for bandits and other actors to declare their tough resistance to authority,[18] as exemplified in the famous novel *The Water Margin* (Shuihu zhuan) in the fourteenth century.[19] In that novel, 108 heroes combat evil, help the poor, and establish their own court to declare noncooperation with the imperial court, although their leader hadn't completely given up the wish to serve the emperor. More broadly, *jianghu* is used to represent "every 'marginal' and dispossessed element in society" (Minford 1997: xxix), such as hermits, performers, knights-errant, beggars, bandits, fortune-tellers, secret-society members, and swindlers (Minford 1997; Liu Yanwu 2003). The nonmainstream space of *jianghu* contrasts with the mainstream space of the court.

The above themes have been reinforced since the 1950s, when martial arts fiction again flourished in Hong Kong, with Jin Yong (Louis Cha) as the most prominent writer. In his novels, the knight-errant is the dominant character in *jianghu*, but the boundary between knight-errant and hermit is sometimes blurred. The knights-errant who carry out heroic deeds often wish to retreat from the chaotic *jianghu* itself. Sometimes, a seemingly quiet hermit is actually a knight-errant with advanced martial arts skills.

The term *jianghu* is also increasingly being used in contemporary narratives. In his documentary *Jianghu*, which records the life of a song and dance troupe, the film director Wu Wenguang (1999) treats *jianghu* as a contrast to home. According to his commentary, the singers and dancers were forced to leave their homeland and were "floating" in another kind of life, full of risks and with obscure prospects.[20] In this example, *jianghu* represents a wandering space for ordinary people who leave their real native lands. "Old *jianghu*" (*lao jianghu*) has also become a common saying, referring to a person who is worldly-wise and adept at dealing with complex situations.

The idea of *jianghu* as a metaphor for the space occupied by Puer tea actors emerged from many scenes that I witnessed during my fieldwork. Both on tea-growing mountains and around tea tables, people "fought" to use open or secret methods to authenticate the tea or person in front of them. I couldn't help associating this with contests in martial arts fiction, such as those acted out by Jackie Chan or Bruce Lee, and the *jianghu* battle depicted in the movie *Crouching Tiger, Hidden Dragon,* directed by Ang Lee, even though the tea competitions were much more sedate. Many people said they were tired of the debates surrounding Puer tea and wished to just sit down for a simple, authentic cup of tea. This reminded me of the hero played by Chow Yun-fat in *Crouching Tiger, Hidden Dragon,* who wishes to give up his warrior life and live peacefully with his beloved.

I found in these scenes intrinsic features of *jianghu* that are applicable to the world of Puer tea. First, as a space located between utopia and reality, *jianghu* represents a social world in which knights-errant achieve romantic dreams, but chaos and risks still remain. Cheating, poisoning, and robbery are frequent occurrences in the world of *jianghu.* Cheating and forgery are also prevalent in the case of Puer tea, making the quest for "authentic" Puer tea difficult to accomplish. Just like the adventures of knights-errant living in the *jianghu,* or of ordinary people leaving their homeland, the route to discovering authentic Puer tea is often full of risk and competition, due to the complex image of Puer tea and the obscure relationships between people. The wish for a simple cup of tea reflects people's anxiety about authenticity and indicates a desire to retreat from these risks and competitions and become a hermit-like tea drinker.

Second, in the *jianghu* world of martial arts fiction, knights-errant could go beyond governmental influence to seek a simple and "perfect" resolution for all kinds of problems: good or evil, right or wrong, all could be judged by matching the *kungfu* skills of opponents. That is, by transcending the social hierarchy, one could decide one's own fate through reliance on personal skill, an ideal many Chinese dream of when they lose faith in formal authority (Chen Pingyuan 1997, 2002). Similarly, facing the lack of clear government regulations on Puer tea, some people believe that the best way to assure authenticity and avoid being cheated is via direct and personal tasting. This is especially evident by the many tea producers, drinkers, and traders who have emerged in the contest over identifying Puer tea's authenticity, who are each proud of their own ability to judge authentic Puer tea. A trader from Guangdong once said to me:

> You don't need to know exactly in what way a piece of Puer tea is planted, processed, or traded. All can be discovered at the tea table. Whether or not too

much fertilizer has been used, what quality of tea the material is made of, which tea mountain it is from, and roughly how old it is can all be determined by your tasting ability.

Thus the spaces for Puer tea actors—whether on faraway tea mountains or in urban teahouses or in online discussions on websites devoted to tea—share similarities with the *jianghu* world for knights-errant. Both contrast with the more standardized rules of the court or the elite in their emphasis on personal skills. In Puer tea's case, this is a subtle kind of noncooperation rather than a clear declaration of resistance to authority. These actors are opposed mostly to complex doctrines or definitions of Puer tea that are declared to be authoritative and scientific but in their opinion fail to distinguish the authentic from the fake.

Third, the term *jianghu* hints that the essence of society is based on the presence of various groups or clans whose disciplines are in debate and cannot be tolerated by one another. That is, each group has its own social space, its "own code of conduct," and its "own language and wisdom" (Minford 1997: xxix). One would be at risk or would lose one's sense of belonging if one entered the other's space without accepting their discipline. In the case of Puer tea, different groups have different palates; divergence exists between the so-called raw Puer tea group (*sheng cha pai*) and the artificially fermented Puer tea group (*shu cha pai*), the Yiwu flavor group and the Menghai flavor group, the Yunnan storage group and the Guangdong storage group. Each declares itself the most authentic and does not tolerate the other. This kind of taste divergence may also be seen in consumption choices discussed by Western scholars over food or clothes that act as distinct cultural markers (see Bourdieu 1984). What is different about the Chinese case might be that the *jianghu* distinction doesn't have a strong emphasis on class hierarchy. James Liu has pointed out that Chinese knights-errant should not be regarded as a distinct social class, "but simply as men of strongly individualistic temperament" (1967: 4). Similarly, Puer tea actors are distinguished not necessarily by class, but on the basis of different interests, mutual prejudice, and indignation about certain "authoritative" instructions. Moreover, in *jianghu* there are generally two levels of distinction: a distinction from authoritative powers that arises by declaring themselves to be a nonmainstream group, and a distinction among themselves that arises from interior discrepancies in terms of different opinions among different disciplines. Even within one discipline suspicions exist, and there are open or secret battles.

The *jianghu* metaphor used in this book is conspicuously embodied by but not limited to the world and behavior of knights-errant that stresses the ethics of loyalty, righteousness, and male camaraderie. Rather, a broad sense of *jianghu* is used to refer to a kind of "intellectual pursuit" by various Chinese—in particular the pursuit for truth, freedom, and independence (*Sanlian Life Week* 2012: 62) reflected by many Puer tea actors' ideal of finding and having a cup of authentic tea.

However, multiple voices have converged as a result of Puer tea's popularity, and counterforces exist between these voices. This makes the effort of "unifying" the *jianghu* of Puer tea (*yi tong jianghu*)—namely, giving Puer tea a singular authenticity—ultimately unsuccessful. Such multiple voices and counterforces have long existed in China. This book both explores this social and cultural continuity and displays the unprecedented depth of a certain commodity's contested fashioning in the context of Reform China's consumption revolution.

The *jianghu* voices and counterforces are illustrated at various spatial levels and in different circles of actors who debate and contest as well as negotiate and cooperate in identifying authentic Puer tea. Generally, this can be seen as a battle between the voices that attempt to package Puer tea within a "perfect" profile, especially for symbolizing certain forms of local nationalism, and the counterforces that unpack its popularity and deconstruct its "perfect" meanings. The measuring of Puer tea's authenticity may take place at a small production place between a local tea producer and an outside trader, or at a consumption site among several tea drinkers. Sometimes the *jianghu* narrative moves to a higher geographical level, with multiple subdistricts in Yunnan competing for the honor of being the "original" producer of Puer tea, or Yunnan itself debating other consumption regions about whose Puer tea tradition is more authentic.

These multiple voices and counterforces are situated in a context where the state has joined in the authenticating process but hasn't been able to efficiently supervise all aspects of production and consumption.[21] According to some producers and traders, this has created added burdens and risks in the *jianghu* of Puer tea. As a result of the state's inability to clearly regulate Puer tea, an arena has developed in which multiple *jianghu* actors express their own voices and find their own solutions, as do the knights-errant who rely on their personal martial arts abilities to survive in the *jianghu*.

China's gift economy is relevant in this regard (Yang 1988, 1989, 1994). Rather than simply reflecting bureaucratic corruption, since the 1980s, the gift economy has actually become an alternative means of distribution and even an important supplement to the state's distributive and redistributive economy. Socializing through gift exchange—known as *guanxi*—may be understood as an unofficial order that arises from the "popular realm" (*minjian*) and is both oppositional and complementary to that of the state (Yang 1994).[22] Likewise, many attempts by ordinary tea producers, traders, and consumers to define authentic Puer tea have come to supplement, as well as subtly resist, governmental regulation.

Although these *jianghu* actors also appeal to the state for clearer and stronger regulation, in actuality, once regulations are clarified, alternative voices emerge to draw Puer tea back to its vague state. The key excuse for retaining the vagueness, I have come to understand from many tea producers, traders, and drinkers during my fieldwork, is that only one's personal experiences—especially sensory techniques—can be relied upon to differentiate quality; scientific standards cannot contend with Puer tea's changeable features and only make the process of tea appreciation boring. Therefore, the problem is not just that governmental regulation is not clear or strong enough, but that this tea culture may not need such clear and strong standardization. Preference for nonstandardization is most obviously exemplified by the supposedly strictest standards on Puer tea's authenticity that are agreed upon by many *jianghu* actors: that authentic Puer tea is the raw kind made with tea leaves from a single origin, from a good environment, crafted by hand, traded via negotiation with small-scale family units, and stored for a long period of natural fermentation. According to this view, the authenticity of Puer tea does not come from mechanization or standardization, but should be judged personally, contextually, and flexibly, and thus it represents *handcrafted authenticity*. It is empirical, flexible, and also vague.

While examining the process of how a substance is fashioned, this book suggests that it is more important to look at the other side—namely, how its fashioned value is counterpackaged by multiple forces. The multiple voices, counterforces, and "handcrafted" standard on authenticity are shaped by the intrinsic features of Chinese *jianghu* culture and are significant in understanding the chaotic phenomenon of Puer tea in the early twenty-first century in China. The intrinsic features of *jianghu* culture shed light not only on consumer culture and business practices in contemporary China but also on the relationship between standardization and individuality, the force of strengthening as well as breaking the social network, and the intellectual pursuit of authenticity by Chinese individuals. This book demonstrates the usefulness of approaches in anthropology, cultural studies, and social science more generally that explore the meanings of consumer goods by popularly relating them to complex aspects of globalization, (post)modernization, state control, transnational regulatory regimes, and reinvented localities. More importantly, this book seeks to go beyond these popular approaches by using unique Chinese concepts, in particular *jianghu,* to interpret how the meanings of consumer goods are deeply rooted in and essentially shaped by accompanying cultural and national characteristics.

A popular question and answer about *jianghu* best explain how handcrafted authenticity is produced: Where is *jianghu* located? In the human heart (*Jianghu zai nail? Jianghu zai renxin*). Here, *jianghu* refers to the space of authentic things. In the case of Puer tea, whether a piece of tea is authentic or not depends upon whether the seller has a kind heart, whether the buyer thinks the tea is good, and the strength of the relationship between the buyer and the seller.

春 SPRING 生

Spring tea buds in Yiwu.
Photo by the author.

CHAPTER 1

"The Authentic Tea Mountain Yiwu"

Our Puer tea is made from tender fragrant tea leaves, topped with the refined tea buds of the authentic tea mountain Yiwu. . . . Recently, fake Puer tea has emerged, counterfeits are increasing, and fake tea is being mixed with authentic tea, which is hard to identify. Some shameless people have counterfeited our brand for their own profit. To forestall such bad effects, we changed our brand icon to two lions in August 1920. Please make note of our special description and avoid being cheated.

—Tongqing Hao Puer tea label, Yiwu, 1920s or 1930s

Our tea boasts original flavor, original taste, and is from an original environment. It is processed by hand and is a healthy green drink. After natural fermentation, it will become mellower: the longer the storage, the better the taste.

—Zheng family Puer tea label, Yiwu, 2007

In January 2007 I passed through Hong Kong on my way from Australia to Yunnan. I decided to stay in Hong Kong for a few days, as I had been told that Puer tea is "produced in Yunnan, stockpiled in Hong Kong, and collected in Taiwan."

While I was in Hong Kong, many people recommended that I go to a famous restaurant, Lianxiang Lou, for *yum cha,* a traditional type of Cantonese cuisine consisting of dim sum and tea and usually eaten for breakfast. Most of the customers there chose to drink Puer tea. Over *shaomai,* or steamed pork dumplings, Zongming, a local Cantonese fellow, said to me, "Now you can understand why Hong Kong people must drink Puer tea. If they didn't, these foods would be too greasy to digest. With Puer tea, people can eat more and stay here longer."

Some days later, Eddie, a local friend, showed me his personal stash of Puer tea produced in different periods. I was surprised to find that he had

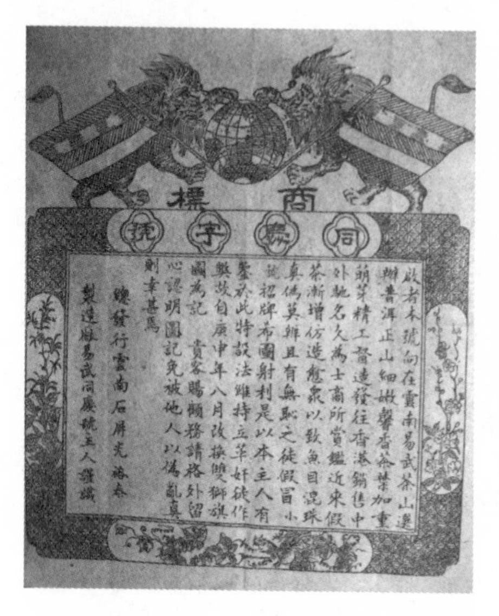

FIG. 1.1 *(left)* Tongqing Puer tea label.

FIG. 1.2 *(above)* Songpin Puer tea label. Photos by the author.

two round cakes of well-known, high-quality aged Puer tea: the first, labeled Tongqing Hao (*hao* means "brand"), had two lions as its icon; the second, Songpin Hao, had its own good luck picture (figs. 1.1–2). I had seen these types only in collectors' books about aged Puer tea (see Deng Shihai 2004: 325–326). Eddie brought samples of both kinds and accompanied me to a famous teahouse in Hong Kong. There we tasted the teas with Mr. Ye, the master of the teahouse, who had a rich knowledge of tea.

These teas were highly appreciated at the teahouse. The waitresses paused in their work and gathered around. Mr. Ye, who had tasted many kinds of aged Puer tea before, infused the tea himself very carefully. He commented that there was a sort of sweetness bubbling up from the depth of his throat soon after tasting the Tongqing tea—like sugar, though more natural. He described the tea's flavor as "like a blossom in the mouth." But he noted that it was still a little acerbic and would benefit from additional aging. He gave higher praise to the Songpin tea, which, he said, fit the model of excellent tea described by an ancient food commentator: it eliminates your arrogance, removes your impatience, elevates your mood, and softens your temper (*ping jin shi zao; yi qing yue xing*) (Yuan Mei 1792).

We talked about the teas' history. Eddie told us that he had gotten both kinds from an old Cantonese man who had bought the cakes early in his life (Eddie was not sure exactly when) in Guangdong, before his family migrated

FIG. 1.3 Aged Red Mark (Hong Yin) Puer tea. The marked prices are in Singapore dollars, for tea on sale in a store in Singapore in December 2007. The tea, according to Puer tea guidebooks, was at least forty to sixty years old. Photo by the author.

from Hong Kong to Australia. Eddie met the old man in Sydney in 1997, and the man gave him several cakes of the tea, calling them "useless relics."

Although it was unclear in exactly which year the tea was produced, Eddie and Mr. Ye dated it to no later than the mid-1930s. That is, both pieces of tea were at least seventy years old. Considering their mature flavor and trademark, they probably came from Yiwu, one of the "Six Great Tea Mountains" in Xishuangbanna, Yunnan. Such aged Puer tea, however, never appeared in my later fieldwork in Yiwu. The tea producers of seventy years ago had good brand awareness, and they used special icons, such as the lions and the good luck motif, to identify the products their family had made in a particular period and to distinguish their tea from counterfeits. Furthermore, these icons were used to represent Puer tea originating from "the authentic tea mountain Yiwu," or *Yiwu zheng shan* (*zheng* means "authentic" or "original"; *shan* means "mountain"). The tea production master included the Chinese character *zheng* in the description of the tea to remind consumers of its authenticity.

In the ten years prior to my visit, the price of aged Puer tea had skyrocketed. I was told by informants in Hong Kong and Taiwan that in the 1980s, when connoisseurs began noticing Puer's value, one round cake (usually 357 grams) sold for less than ¥1,000; its value then reached over ¥10,000 in the 1990s, and in 2002 a single cake was traded for more than ¥1 million. It's no surprise that some tea connoisseurs like to say that "one cake consumed is one cake lost" (see also fig. 1.3). Such aged and valuable Puer tea cakes were established as ideal specimens, inspiring connoisseurs to make a pilgrimage to their place of origin, Yiwu. As with some other commodities, such as French wine (Ulin 1996; Guy 2003) and chocolate (Terrio 2005), Puer tea's burgeoning popularity relied heavily on the role of connoisseurship. The connoisseurs created a series of standards for authenticating Yiwu's raw tea, which were further developed by traders, tourists, writers, media reporters, and locals, who together packaged an "authentic" image of Yiwu.

HISTORICAL GLORY

Yiwu is a township in Mengla County, Xishuangbanna Autonomous Prefecture, in southern Yunnan (map I.2). It is located on the east bank of the Mekong River, close to the border with Laos. The township covers a mountainous region of 864 square kilometers. Its population of about thirteen thousand is composed of 34 percent Han and 66 percent other ethnic groups, of which 29 percent are officially recognized as Yi, 21 percent as Yao, and 16 percent as Dai (YTG 2007).

Prior to tasting the aged Puer tea in Hong Kong, I had been to Yiwu twice. The first time was in 2002, when I happened to be there with a crew that was making a film about people involved with the tea business in Yunnan. On that trip, I stayed for only one day. The second time was in early 2006, when I went for one week, with a clearer aim to observe the area for future fieldwork. I was struck during those early journeys by two things: the unique way that tea makers shaped Puer tea into cakes using their hands and the weight of a stone press, and the typical Han style of old family houses surrounded by rural scenes. As it turned out, these two features were closely linked, because the tea is processed in the old houses.

I learned more about the tea from Zhang Yi, the retired leader of Yiwu Township, who had taken the lead in reviving the technique of handcrafting Puer tea in the mid-1990s, after Puer tea had gone largely unnoticed in Yiwu for almost half a century.

Zhang had worked on local chronicles and was so familiar with the history of Yiwu that in 2002 he was invited by the film crew to act as a guide. Following him into the old houses, I heard stories about families whose histories were bound up with the Puer tea production and trade of many years before. These stories were recounted by Mr. Zhang and other locals in the Shiping accent, a Yunnanese dialect. According to their versions of local history, from the mid-seventeenth to the eighteenth century, many Han migrated to Yiwu and the nearby regions from Shiping, a southeastern county in Yunnan (Zhang Yi 2006b: 72–73; see also Zhang Yingpei 2006: 77). Before the arrival of the Han, the indigenous people (*ben ren*) had cultivated tea trees, usually under the authority of Dai overlords. At present, these indigenes are identified by the authorities as Yi. Some researchers think that this ethnic group is actually closer to Bulang (Mu Jihong 2004: 63), while others argue that it is closer to Hani (Gao Fachang 2009: 29). In reality, many non-Han people in the area describe their ancestors as indigenous, but call themselves Yi in accordance with the official classification.[1]

According to one study, before the Han arrived, at least 5,000 *mu* (a unit of measure equivalent to 0.0667 hectares) of tea lands were cultivated by the indigenous ethnic group (Zhang Yingpei 2006: 75–77). The new Han immigrants gradually acquired rights to the land, sometimes by buying it and sometimes through intermarriage with local Dai aristocrats (Dao Yongming 1983: 61; Liu Minjiang 1983: 57–58; Hill 1989: 332). By making use of obsolete areas and clearing new areas, they developed their own tea plantations and came to rely on tea as an important part of their economy.[2] As a result of the efforts of both the Han migrants and the original groups, the tea cultivation area increased. The loose tea would undergo rough processing by the tea growers before being traded to the Han merchants, who established commercial tea companies that organized fine processing for pressing the tea into cakes and traded the products (Liu Minjiang 1983: 57; Zhang Shixun, interview, 2007). According to recent research by Puer tea experts based upon examining examples of old tea and the genealogies of tea producers, many commercial brands, such as Tongqing Hao, were established by Han families in Yiwu in the eighteenth or nineteenth centuries (Deng Shihai 2004: 82; Zhang Yingpei 2006: 122–123). Puer tea made in Yiwu and the nearby mountains entered a buoyant commercial period and became very famous. The region prospered, and according to one writer from 1799, "hundreds of thousands of people came to trade in the tea mountains" (Tan Cui [1799] 1981: 387). The trading networks of Yiwu's Puer tea extended in several important directions: to Beijing, Southeast Asia, Hong Kong, and Tibet (see map I.1).[3]

Now, after many years of wear, old family houses that had once been famous commercial tea companies stand along the old street of Yiwu. Some are still occupied by the families' descendants, others have been transferred to new owners, and still others have been destroyed, with only the stories told about them remaining. The houses of Tongqing and Songpin, whose tea I encountered in Hong Kong, have disappeared.

When I was with the film crew in 2002, our guide, Zhang Yi, took us to one particularly well-preserved house, which used to be the site of the famous company Cheshun Hao. From some rough tea material piled on the ground of the parlor, the owner took out an old dusty wooden signboard. The film crew took a shot of the board, for which they were charged ¥10 by the master, a descendant of Cheshun Hao. The signboard was inscribed with the words "Tribute to the Emperor" (*Rui gong tian chao*); in return for sending Puer tea to the emperor of the Qing dynasty (1644–1911), the emperor granted them the signboard. As Zhang Yi and other locals pointed out (fig. 1.4), this was the highest possible honor.

There are six connected tea areas east of the Mekong River in Xishuangbanna: Yiwu, Yibang, Manzhuan, Gedeng, Mangzhi, and Youle.[4] Collectively, they are known as the Six Great Tea Mountains (Liu Da Cha Shan). All are in Mengla County except for Youle, which is in Jinghong City. The fame of these tea mountains is linked to the tribute tea sent to the Qing emperor in Beijing during the eighteenth and nineteenth centuries. The Six Great Tea Mountains used to be controlled by the Dai ruler of Jinghong, the indigenous regime in Xishuangbanna. Later, the Qing administration gradually extended its control into this area, conquering local officials and replacing them with imperial bureaucrats, under a political reform known as *gai tu gui liu* (literally "replacing *tusi* [native officials] with imperial officials"). A new Puer Prefecture of Qing China was established in 1729, with Puer as its capital. At that time, the east bank of the Mekong was removed from Dai control and added to Puer Prefecture. Shortly after, the Six Great Tea Mountains became the base for tribute tea to Beijing (1732–1904) (Giersch 2006; Zhang Yingpei 2006). The oldest known cake of Puer tea, Golden Melon Tribute tea (*jin gua gong cha*), found in the Palace Museum in Beijing (now kept in the Chinese Tea Institution in Hangzhou) (see the imitation in fig. 3.3), raised the profile of the Six Great Tea Mountains, because the tea material originated in Yibang, which was also inhabited by indigenes, Dai, and Han migrants (Zhang Yingpei 2006: 109). Golden Melon Tribute tea was produced more than one hundred years ago. Yibang was the politi-

FIG. 1.4 The entrance to Cheshun Hao, 2007. Photo by Sun Jingfeng.

cal, administrative, and tea-trading center of the Six Great Tea Mountains from the 1750s until the early 1900s. After Yibang's decline, Yiwu rose to prominence in the early 1900s and became the production and distribution center for Puer tea (Zhang Yingpei 2006: 11–14).

A common saying is used to demonstrate the Qing royal court's appreciation of Puer tea in their daily diet: "Dragon Well tea (Longjing) for summer, and Puer for winter" (Huang Guishu 2005: 86–88). Among the various types of Chinese tea, Puer is thought to be the most helpful in digesting greasy food. This feature is said to have perfectly suited the needs of the Qing royal families, who were descended from the northern nomads of China and had meat as their staple food. After an initial rough processing in the Six Great Tea Mountains, the tea materials were sent to the capital, Puer, or to Simao, where the General Tea Bureau (Zong Cha Dian) was established, for fine processing.[5] This fine processing took place under strict supervision, and the finished products were carried overland to Beijing.

Apart from its history of providing tribute tea for Beijing, Yiwu is considered unique today because of its valuable aged Puer teas, such as Tongqing Hao and Songpin Hao, which are collected by connoisseurs in Hong Kong and Taiwan. These teas are said to be the oldest examples of Puer tea aside from the Golden Melon. In the nineteenth and early twentieth centuries, before modern transportation in inland China, Puer tea originating in Yiwu and the nearby mountains was sent by caravan to Lai Châu (now in Vietnam) or Phongsali (now in Laos), and then farther afield to Southeast Asian ports like Hải Phòng and Bangkok.[6] Finally, it was transported by boat to Hong Kong. The overland trade and transport was dominated by Han traders from Yiwu and the nearby tea mountains,[7] and the overseas part was often undertaken by Cantonese traders originating in Guangdong (Prasertkul 1989: 51, 73–74; Luo Qun 2004: 244; Zou Jiaju 2005: 57; Zhang Yingpei 2006: 83). Young Puer tea thus flowed to Hong Kong and was aged and accumulated there. Cantonese people say that in Hong Kong, even children would be encouraged by their parents to drink Puer tea while having *yum cha* (Chan 2008: iv).

Puer tea was also sought after to balance greasy food in Tibet, which has long been regarded as the most important buyer of Yunnan's tea. Tibetans are said to have started drinking tea, which they imported from Yunnan or Sichuan, in the eighth century (or, by one account, the third to fourth century).[8] More tea was traded from southern Yunnan to Tibet during the Ming dynasty (1368–1644), and the tea trade flourished under the Qing (1644–1911). The Qing government began issuing tea certificates to Tibetan merchants at Yongsheng in 1661, and at Lijiang starting in 1748. After receiving these certificates, the Tibetan traders could go to Puer for further tea trading (Fang Guoyu 2001: 429; Zhou Hongjie 2004: 4).[9] This trade continued during the Republican period (1912–1949), but it was often blocked by wars and political conflict. In Yiwu I heard an interesting story about the Tibetan trade from Zhang Yi, as well as from some other elderly people. In 1945, a group of Tibetans arrived and impressed the local Yiwunese with their tall figures. They bought all the tea they could find in Yiwu—even the tea stored in the henhouse, according to one exaggerated account. The Tibetans had thirsted for Puer tea for many years during the Second World War, when the caravan routes were blocked, and they traveled a long way to buy the tea even though some of their horses had died on the way to Yiwu.

The Puer tea business in Yiwu was not always successful. It experienced several periods of decline, mainly due to political factors. During World War II and the civil war between the Nationalists and the Communist Party in the 1940s, the tea trade virtually came to a halt. A few years after the foundation of the People's Republic of China (1949), the purchasing and selling of foodstuffs, including tea, was monopolized by the state, and the operation of private family business was completely stopped. From the 1950s to the 1990s, Yiwu and the nearby tea mountains produced basic tea material primarily for state-owned tea factories (Zhang Yi 2006b: 34).

Though political struggle became the dominant theme in the Maoist era, from the 1950s to the late 1970s, and many activities were sacrificed in the name of class struggle, Yunnan's tea production didn't stop. State-owned tea factories kept producing tea—to supply the needs of Tibet, on the one hand, and to supply Hong Kong, Macao, and other countries in Southeast Asia and Europe, on the other. Maintaining a supply to Tibet was considered important for interethnic relations, while supply to Hong Kong and other countries was a way of gaining foreign currency (YTIEC 1993: 7–11, 160–165).

When Hong Kong returned to mainland China from British rule in 1997, many Hong Kong residents, worrying about the political change, moved overseas and sold off their Puer tea stocks, which they had collected over many years. The biggest buyer was Taiwan (Deng Shihai 2004: 84). Ironically, it was thus not people from Yunnan, Beijing, Tibet, or even Hong Kong who launched the rediscovery of the origin of aged Puer tea, but a group of Taiwanese "tea madmen" (Zeng Zhixian 2001: 109).

In October 2007, in Yiwu, I met Mr. Lü, the head of that pioneering Taiwanese group. He told me that he had brought aged Tongqing Puer tea with him when he and his group first came to Yiwu in 1994, as he was determined to obtain more of this kind of tea. However, when he expressed this desire in Jinghong (the capital of Xishuangbanna), the tour guide responded that she was unfamiliar with this place called "Yiwu" and wondered why it was necessary to visit it. Lü and his group insisted that they knew from historical accounts how important Yiwu had been for Puer tea. But at the time, the artificially fermented tea was more commonly recognized as Puer tea, so Mr. Lü and his group were told by government officials in Kunming, Simao, and Jinghong not to go to Yiwu because Yiwu didn't produce artificially fermented Puer. But Mr. Lü and his group knew that the valuable aged Puer teas, like Tongqing and Songpin, were naturally fermented from the raw tea of the Six Great Tea Mountains.

They were deeply frustrated when they finally reached Yiwu. They found neither aged Puer tea in storage nor new caked Puer tea in production. People continued to produce basic tea material in loose form (rough processing) in order to dispatch it to national tea factories, but the technique of making round-caked Puer tea (fine processing) had been lost. Lü told me that what they saw was only a ruined and depressed Yiwu: bad transportation, few restaurants, a simple and crude guesthouse, old family houses, and village roads in disrepair. No one knew anything about aged Puer tea. Finding that the Puer tea industry had been inactive there for almost half a century, Mr. Lü and his group tried their best to focus on a few bright points: First, the tea resources were still there. Remarkably, the old and tall tea trees within the forest were still being cultivated by both ethnic minorities and Han. Second, some older people who had worked on caked Puer tea might still be alive in Yiwu. Third, aged Puer tea originating from Yiwu many years ago, which the Taiwanese brought with them, could be used as a model in reviving the industry. Although the basic tea material in loose form was recognized by Mr. Lü and his group as raw Puer tea, it was not convenient for transport, and all the valuable aged Puer tea had long been stored in pressed cake form.

Mr. Zhao, who worked in the government office of Yiwu Township and participated in the reception for the Taiwanese in 1994, recalled his astonishment:

> We were completely surprised at suddenly having a group of Taiwanese guests, and we could not understand why they took the trouble of traveling to our rough, rural, and remote small village. Mr. Lü, the head of their group, invited us to share the aged tea he brought with him, and we later learned that it was Tongqing Hao. I thought its flavor was great, and it had a special smoothness, which I had never experienced with tea. He asked us to guess the price of this tea in the external market. I daringly estimated "¥400 or ¥500 for one round cake," thinking to myself that this might have been too much. However, we could not believe our ears when Mr. Lü said that it was sold for ¥15,000 a cake in Taiwan!

Finally the Yiwu government decided that the wishes of the Taiwanese guests should be satisfied as much as possible. They recognized that the Taiwanese had made a long journey in pursuit of Puer tea. They also recognized that the relationship between mainland China and Taiwan was difficult but improving, and they were excited about the value of Yiwu's Puer tea in the outside world.

Two men in their sixties, who had been hired as workers for Tongqing Hao before 1949, were sought out and invited to be the new tea masters. Using the model of the aged tea cake brought by the Taiwanese, a process of teaching, learning, and imitating began. Zhang Yi, who at that time was working on the local chronicles, participated in the tea-processing classes and later developed a business relationship with the Taiwanese visitors. Thus developed the revitalized process of Puer tea making that was shown to the film crew eight years later, in 2002.

TASTE PREFERENCES

In his clean yard Zhang Yi put some loose tea material into an iron cylinder, steamed it for about ten seconds, and then poured it into a cloth bag to shape with his hands into a round cake. The caked tea was then pressed under a stone press, with someone standing on it to add weight. After being put on a wooden shelf for several hours, the tea was taken out of the cloth bag for further drying. Finally, seven pieces were wrapped together with bamboo leaves in a stack, known as "seven-son tea cake" (*qi zi bing*) (fig. 1.5).

This fully manual way of processing tea is practiced in Yiwu and the Six Great Tea Mountains, all located to the east of the Mekong in Xishuangbanna. By contrast, a mechanical method is commonly used in Menghai, another district in Xishuangbanna, to the west of the Mekong River (map I.2). There are other interesting contrasts between Yiwu and Menghai, with the river as the boundary between them.

The authentic image of Yiwu and its Puer tea is constructed by reference to "the other," not just by reference to the past. That is, its identity is constructed not only by looking at Yiwu itself, but by juxtaposing, competing with, and representing the other (Baumann 1992; Ohnuki-Tierney 1993).

In terms of tea resources, both Yiwu and Menghai have tall tea trees in the forest as well as the "bush" form of terraced tea fields, though their different soil and climate lead to different tea flavors. For rough processing, they share similar procedures in tea harvesting, killing the green, rolling, and drying. A greater contrast lies in their techniques of fine processing, which have been shaped by the different ways the tea industry has developed in the past.

When the tea business in the Six Great Tea Mountains prospered during the Qing dynasty, from the mid-eighteenth to the early twentieth century, Menghai was less famous for Puer tea, even though it also possessed excellent tea resources. However, the situation gradually changed, and Menghai's geographical advantage became more and more prominent, since it borders

FIG. 1.5 Caked teas just shaped by a stone press. Photo by Sun Jingfeng.

Burma and is located in the lowland plains, which were more suitable for modern transportation. A turning point—which finally broke through the dominant position of the Six Great Tea Mountains in Puer tea production and trade in Xishuangbanna—occurred in 1938, when the first tea factory of Yunnan, the National Tea Factory of Menghai, was opened. This was a branch of the Central Tea Company of the Republic of China established by the national government for the purpose of improving red tea production and export. Advanced machines brought from British India and Burma began working on processing tea in the Menghai Tea Factory, marking the start of mechanical tea processing in Yunnan. But Yiwu maintained its handcrafted technique. This difference in processing techniques persists today. Different terms are used to describe production units in the two places. For Menghai, the term is "tea factory" (*chachang*); in Yiwu, it is "household tea unit" (*chazhung*). The former stands for the rapid and large-scale *modern* production, while the latter persists in slow and small-scale *traditional* production.[10] After 1938, the private tea business of Yiwu was knocked back by the policy support for the national factory in Menghai; combined with the other events mentioned earlier, Yiwu's dominance gradually waned (Zhang Yingpei 2007: 39–40).

Another turning point occurred in 1973, when artificial fermentation was invented in Kunming in order to satisfy some consumers' wish to produce a mature tea in a shorter time. The technique was applied in Menghai soon after its invention, and the result, known later as artificially fermented Puer tea (*shu cha*), could be finished in two or three months. Today Menghai is known as the best place to produce artificially fermented Puer tea due to its special natural environment as well as its advanced techniques. "Menghai taste" refers to the unique flavor of Menghai's artificially fermented Puer tea.

This artificial fermentation is more technically complicated than producing raw Puer tea. A national tea factory was never established in Yiwu to popularize artificial fermentation. In fact, many Yiwu people despised this processing technique and preferred not to drink the resulting tea. The iconic Puer tea of Yiwu had always been raw tea (*sheng cha*) intended for natural fermentation. I asked many people in Yiwu why they didn't produce artificially fermented Puer tea. Some replied, "That's not our tradition," or "We don't know how to do that." This shows that artificial fermentation belongs in the large modern tea factory rather than in the small traditional home production unit. "Using Yiwu tea material for artificial fermentation," one young tea producer remarked, "is like burning a pile of money," indicating the pride of Yiwu people in their excellent tea resources. In their minds, high-quality tea material is good enough to drink after simple processing and should only be fermented naturally. One female tea producer told me, "I've witnessed the process of artificial fermentation. It's not clean at all. Is that kind of tea drinkable?" Her concerns were echoed by other Yiwunese, who regarded the fungi that grew on the tea in the process of artificial fermentation as unclean and harmful.[11]

These preferences were rooted in Yiwu's production tradition and affected by the authenticity standard that had recently been initiated by Taiwanese connoisseurs and further developed by other traders and consumers. As several Taiwanese traders explained to me, the rare and valuable Tongqing and Songpin brands, and other aged teas of the same generation, were all naturally developed from raw material originating in Yiwu and the nearby mountains. That is, the Taiwanese preference "strongly emphasized the concept of originality" (Yu Shuenn-der 2010: 133). They also contended that handcrafting would make the Puer tea cake sufficiently "loose" to be suitable for natural fermentation, whereas Menghai's mechanical processing led to a hard "discus" of tea that was too tightly packed to allow contact with the air. Such distinctions were elevated by some Puer tea writers, from both

Taiwan and Yunnan, to a more abstract level. They argued that handcrafting and natural fermentation stand for superior culture and embody the essence of Daoism (Lei Pingyang 2000; Deng Shihai 2004).

In terms of the binary contrast between "the raw and the cooked" (Lévi-Strauss 1970; 2008; see also Leach 1970), as artificial interference increases, food moves further from nature and closer to culture. Handcrafting involves more human labor than mechanical production does; therefore, handcrafted tea is further from nature and closer to culture. But in terms of the naturalism of the raw Puer group, this is only the first step. They also believe that tea aged naturally produces the purest taste and superior spiritual enjoyment (Deng Shihai 2004; Lei Pingyang 2000). This pursuit is rooted in traditional Chinese philosophies, such as Daoism, which look at human beings as part of the unity of the cosmos and seek to follow the way of nature. The art of Japanese food, which should be made "as close as possible to the natural state of the foodstuff" (Ashkenazi and Jacob 2000: 86), also embodies this idea of Daoism. In this sense, the raw and natural essence of Puer tea does not mean that there is no human intervention at all, but that "the human intervention must be as 'natural,' that is, as minimal as possible" (Ashkenazi and Jacob 2000: 86). It shows that culture can be "cooked" not only by removing it from nature, but also by "preparing" it in a way that contains the essence of nature as much as possible. This approach displays the continuity of one important characteristic of Chinese culture in seeking authentic things: it seeks things that look natural but that, nevertheless, must be subtly and artificially processed. It echoes a saying about Chinese arts, such as the Chinese garden: "It is man-made, but it looks like it was created by nature" (*Sui you ren zuo, wan zi tian kai*).

Craft objects may be embedded with more symbolic meanings than mass-produced ones. Craft objects are also more closely linked to localized authenticity and can be used to construct a distinct "cultural authenticity" to counteract the globalization of production and consumption (Terrio 2005). Similarly, the preference for raw tea in Yiwu is contextually selected and utilized by Puer tea actors to construct their ideal nature and culture. To them, artificially fermented tea is technically produced by an unknown other, while raw tea is to be "naturally" stored by oneself. Though such natural storage is also a product of human action, importantly, it is "processed" by the collector. In this regard, the self-collected and self-stored tea becomes even more authentic because its postproduction involves personal participation and therefore is more closely tied to identity. The artificially fermented tea of Menghai lacks this personalized authenticity.

When the outside traders chose the unique handcrafted Puer tea of Yiwu, they were also choosing a unique type of social relationship. They wanted to negotiate with the small-scale family unit rather than the large-scale factory. Some traders described the relationship with the former as "warm" (*nuan*), whereas the relationship with the latter was "cold" (*leng*). Regarding the Chinese construction of social relations, the contrast between an "uncooked and raw person" and a "cooked and ripe person" has been used to stress that only via identity transformation from the former to the latter could an "outsider" be accepted as an "insider," thus bridging the social gap. In Chinese culture, the boundary between self and other is often unclear, and personal identities are not necessarily bound to human nature but "constantly being created, altered, and dismantled in particular social relationships" (Yang 1989: 40).[12] Likewise, traders in Yiwu found that it was easier for them to be transformed from uncooked to cooked when dealing with small-scale family units rather than big factories. Some of them even stayed with the family throughout the tea harvest season, seeking chances to talk and dine with the family members and sometimes even cooperating with them on some aspects of tea production and trade. Through this cooperation, traders made the final tea products more understandable, imbued with more human effort and emotion, and hence more authentic.

In addition to distinguishing between the different production processes, many locals, traders, and consumers drew a contrast in taste between teas from Yiwu and Menghai. Yiwu has always focused on raw Puer tea, but Menghai now produces both raw and artificially fermented Puer tea. In Menghai, all I heard about was the "Menghai taste," which was later extended to cover raw tea as well as artificially fermented tea. When applied to the latter, Menghai taste refers to the mellow flavor of the tea that results from the advanced fermentation technique, and the special climate and water resources that contribute to this process. When applied to raw Puer tea, Menghai taste refers to a strong tea energy (*cha qi* or *ba qi*) and the deep and long-lasting sweetness in the throat that follows the strong and bitter flavor (*huigan*). These features are attributed to the excellent and special tea resources in Menghai. Historically, Yiwu tea has commanded a higher price than Menghai tea, but a particular sub-tea mountain in Menghai, called Laobanzhang (see Banzhang in map I.2), has stood out in recent years, and its tea price has exceeded that of Yiwu. The tea from Laobanzhang has become a superstar, praised as the "king" of all raw Puer teas.

In contrast, Yiwu's raw Puer tea was described in softer terms, though its softness was said to also contain strong features. Yiwu's soil tends to be more acidic and Menghai's more alkaline; objectively, this results in a difference in taste. According to some Yiwu tea lovers, the raw Puer tea of both Laobanzhang and Yiwu is strong, but the strong flavor of Laobanzhang is too direct, whereas the strong flavor of Yiwu is subtle and gradually leads to a lingering aftertaste.

However, in the opinion of those who prefer the Menghai taste, all these soft advantages and culturally significant features became disadvantages. In their view, Yiwu's tea was tasteless, or, in the words of a trader, "only a cultural speculation."

Jian, a trader from Jiangxi (in eastern China) who went to both Menghai and Yiwu, told me that he had considered this distinction for a long time. He observed:

> The Mekong is the boundary. On the west side, tea from Menghai is liked by Hong Kong and Guangdong traders. They prefer strong-tasting tea, which accords with their food habits; they don't pay too much attention to "culture" but just want to acquire more varieties and quantities of tea, which is to the advantage of the tea factories in Menghai. The Six Great Tea Mountains, to the east of the Mekong, belong to Taiwanese traders. They care very much about the "culture" attached to the tea; for example, they practice more classical Chinese arts and have more respect for traditional culture, and therefore are happier about Yiwu tea. They are the opposite of the Cantonese traders. The Taiwanese care more about quality than quantity, which the handcrafted way of making Yiwu tea is suitable for. Therefore, I'd say that Hong Kong and Guangdong traders *collect* tea (*shou cha*), but Taiwan traders *refine* tea (*zuo cha*).

In sum, Menghai, standing on the west Mekong, represents the modern mechanical method: large-scale factory production, accelerated fermentation, and strong "king" taste. To the east of the Mekong, Yiwu stands for the traditional handcrafted method: small-scale family units, natural and slow processing, a soft, gentle, but lingering aftertaste, and with more "culture" elements.

Mentions of Yiwu and the Six Great Tea Mountains in historical records[13] reveal this region's importance in the tea trade, especially during the Qing and early Republican periods, but the records do not contain much detail. By using works such as *Selected Works of Historical Accounts of Xishuangbanna, Volume 4* (Banna wenshi ziliao xuanji di si ji) (Zhao Chunzhou and Zhang Shungao 1988), the Taiwanese pioneers launched their journey as well as their new narrative about Yiwu. Zeng Zhixian, one of the Taiwanese pioneers who came to Yiwu in 1994, published a book in 2001, *Experience within the Micro Circumference: A Deep Exploration into the World of Compressed Tea* (Fangyuan zhi yuan: Shen tan jin ya cha shijie), which was one of the earliest works to depict Yiwu in detail with text and rich illustrations. In Zeng's narration, Yiwu became a space where the past and present converged:

> With the historical record as a reference and with the storytelling about the old teahouses by local elites, in our mind there came prosperous scenes of tea processing surrounded by people coming and going. Looking at aged architecture in disrepair but still representing the classical ancient style, walking along the Ancient Tea-Horse Road, we imagined that the boundary of time was broken, and we were going back to the past during the late Qing and early Republican period, sharing travel together with the famous masters of those well-known tea families. (Zeng Zhixian 2001: 109)

Illustrations show the old architecture, the flagstone path on which the ancient caravan had traveled, the famous stone monument recording a local legal case on tea, and the inscribed signboard. According to Zeng, these are not dead relics; they have witnessed the vicissitudes of the old family tea brands and survive in the livelihoods of their descendants.

Echoing Zeng, the subsequent literature, mass media illustrations, tour designs, and commentaries all reiterated this nostalgic theme, encouraging tourists to imagine the glorious past by using the present relics as reference points. Like many popular narratives in the world that had boosted the popularity of a certain cuisine (see Appadurai 1988; Ferguson 1998) or a certain place (see Ivy 1995; Notar 2006b), this literature played an important role in packaging Yiwu and its Puer tea.

Zhang Yingpei, a journalist from Kunming, has written several well-known works on Yiwu (Zhang Yingpei 2006). When I interviewed her in 2007, she told me that one of her motives for writing about Yiwu was her uneasiness about the fact that the previous works about Yiwu were written mostly by Taiwanese authors. She claimed that in those books she had found some serious mistakes; for instance, they were wrong about the date that Tongqing Hao was established, which would mislead consumers trying to identify an authentic piece of Puer tea. As a Yunnanese, Zhang Yingpei felt that she could do better than the outsiders. Her critical attitude demonstrates a strong identification as Yunnanese, in which conveying accurate information about the production site of Puer tea becomes an essential mission.

Locals also joined in this historical research and writing. Zhang Yi, the retired township head, was famous for this. He was well educated, he had worked on local history in the government, and he had cooperated with the first group of Taiwanese connoisseurs. He claimed to have a rich knowledge of Yiwu and its Puer tea and had become the most important source of firsthand information for many visitors and media workers. The other famous local writer, a middle-school teacher, put his mission directly: "Only Yiwunese can clearly describe Yiwu events."

Local people had increasingly realized the value of Yiwu's past. They welcomed researchers to the town, conveyed stories inherited from their ancestors, showed their own book collections on Yiwu and Puer tea, and suggested that the researchers visit certain places or interview certain people. Many of the locals had gotten used to being disturbed by visitors, and they warmly welcomed these disturbances, believing that historical discovery would boost their tea business.

Puzi (web name), who writes a column on a tea website, was delegated by a tea company to go to Yiwu to write a new book about the Six Great Tea Mountains. As his companion joked, Puzi was always happy when an old man appeared, because he would be able to dig out more stories. He interviewed his sources with detailed questions, as an anthropologist would, as though the tea would be more flavorful if this detailed history could also be infused into the cup.

Traders, whose original task was to trade good tea rather than history, also participated in this research. While collecting tea itself, some traders endeavored to gather interesting stories to share with their future customers. In their teahouses in Kunming, maps of the Six Great Tea Mountains were hung, and pictures of Yiwu's tea trees, old architecture, and handcrafted ways of making tea were displayed. The masters of the teahouses who were able to annotate these images with stories were respected by their clients as perfect tea experts, and the Puer tea infused with this history was endowed with a unique cultural taste, and deemed worthy of being more carefully savored.

Through all these efforts, Yiwu was presented by multiple actors as a tea locale with rich culture (*wenhua*). According to those proponents, first, the culture in Yiwu lies in a series of concrete representations, as illustrated by the pictures in Zeng's book: the old architecture, the remnants of the Tea-Horse Road, and various relics. All these things are taken as living antiquities, silently but persistently showing the ever-prosperous period of Yiwu.

Second, these silent antiquities needed to be awakened and vocalized. Past stories about tea families needed to be dug out, iterated, and verified. In this sense, history, which involves wonderful stories and rich human characters, is equated with "culture." When the historical moment and the present converge in a real place, nostalgia—a necessary emotional experience for visitors—is aroused.

Third, this nostalgia for Yiwu is put in a niche because it is so well linked to its representative commodity: Puer tea. Generally, nostalgia refers to recalling the past with memory. But, in a "further globalizing twist," nostalgia could be "without memory" and used "for the present" (Appadurai 1996: 48; see also Jameson 1983). The past life of the older generation can be shared with and memorized by the younger generation, who have never had the experiences being remembered (Bloch 1998); the overseas Hmong, for example, have a strong nostalgia for their presumed ancestral place in China, a place that they haven't ever visited, and they intend to reunite with "homeland" people whom they have never been separated from (Tapp 2003). The cult status of aged Puer tea evokes a similar nostalgia. In popular writings, aged Puer tea is described as the best tea "bearing the weight of time" (Ruan Dianrong 2005b), symbolizing the hard journey of the old caravans (Zeng Zhixian 2001), representing elders with much experience (Mu Jihong 2004), and bearing the stories and experiences of unknown others because it has been stored and circulated via several hands (Ruan Dianrong 2005b). Buyers of aged Puer are encouraged to look back nostalgically to a world they have never lost. This is nostalgia for "imagined pasts never experienced" (Lowenthal 1985).

In the preference for raw tea in Yiwu, a new version of nostalgia is created. Collecting, storing, and consuming raw Puer tea is part of one's own experience and memory; it starts from the present, and therefore the social life of this Puer tea is clarified by rather than dependent upon mythical and unaddressable others. This can be nostalgia using one's own memory, and the collection and storage of raw tea is preparation for nostalgia in the future, when the raw Puer tea becomes aged. Advocacy for raw Puer tea is thus advocacy for a future that hasn't come, yet has already been designed.

Fourth, the raw Puer tea processed by hand rather than by machine is considered to be imbued more deeply with cultural meanings and a unique 'natural' aura.

Fifth, in China, tea has long been considered more than a simple drink to quench one's thirst. It has been depicted in poems, paintings, and calligraphy, the so-called "elegant activities." Yiwu and its Puer tea, surrounded with more historical records and relics than tea from other areas, such as Menghai, typify this concern. This point was understood even by locals with limited education. For example, Mr. He, a man more than sixty years old, often told his son: "Tea is better than alcohol. With alcohol, people often behave badly; with tea, everyone becomes elegant and refined in manner [*wen zhi bin bin*]."

Visiting Yiwu has become a pilgrimage for people who are eager to know about Puer tea. In the classic tour route, they first pay a visit to the aged tea trees in the forest (fig. 1.6). Although this places more emphasis on nature than culture, it is a necessary part of the tour because it shows visitors the "original environment," where the "original taste" or "original flavor" comes from.

Next they walk around the old street of Yiwu, where the old family houses connected by the remnants of the Tea-Horse Road can be seen. In October 2006 a new museum opened at the site of the original temple, where various relics of Yiwu and the Six Great Tea Mountains have been gathered for display.

Finally, the travelers usually go into some family homes, especially those with attractive architecture. Being recognized as possible traders or clients, or journalists or researchers who have come to discover more about Yiwu, the visitors are usually warmly welcomed by the owner of the house, allowed to have a careful look at the house, and even invited for a Shiping-style meal with homemade soy sauce. Even if meals are not served, Puer

FIG. 1.6 This old tea tree in Yiwu is over 10 meters high; the diameter of the trunk is around 40 centimeters. Some tea experts say it is more than five hundred years old. Photo by the author.

tea always is. If they are lucky, the travelers may chance upon the family processing of caked Puer tea. Travelers are often allowed to stand on top of the stone press and personally experience the way of "tradition." Seeing the family working and being allowed to participate makes the production procedures visible. The original unfamiliarity or "alienation" of travelers to the tea products is largely removed (Terrio 2005: 149), and the production process itself becomes consumable and "iconic of a kind of authenticity" (Dilley 2004: 805).

CONCLUSION: IMAGINED ORIGINALITY

Standards for the authenticity of Yiwu and its Puer tea are constructed by locals and nonlocals alike. On the one hand, in the age of mechanical reproduction, the concept of authenticity is closely tied to "originality," as products regarded as authentic must contain an original "aura" (Benjamin [1936] 1999: 72–79). The standards for authentic Yiwu and authentic Yiwu

Puer tea emphasize this original and unique "aura." The traditional handcrafting of tea has inspired nostalgia and "missions" to discover historical "truth." On the other hand, Puer tea goes beyond singularity, uniqueness, and irreplaceability. When copies of products proliferated as a result of mechanical reproduction—such as print, photography, and film—in the nineteenth century, people began to worry about originality (Benjamin [1936] 1999: 72–79). In the case of Yiwu Puer tea, the original (aged tea and traditional handcrafting) is respected, while reproduction of copies (raw tea) is encouraged and said to feature original handcrafting rather than a mechanical process.

Is raw or aged tea more original and authentic? When raw Puer tea was made in Yiwu fifty years ago, it was not destined to be consumed only after aging, and the value of aged tea wasn't widely advocated until connoisseurs, especially those from Taiwan, began visiting the production regions. Perhaps one could say that raw Puer tea is the original because it is actually the basis of the aged. However, the aged tea exported from Yiwu over fifty years ago and recently brought back by connoisseurs is becoming the standard model. So one could also say that the valuable aged kind is the original and represents the authenticity that connoisseurs seek. Nevertheless, the availability of aged tea is limited and decreasing. In order to procure more "authentic" and valuable products, connoisseurs, together with other actors, deliberately but inconspicuously replaced the object of authenticity, temporarily overlooking aged flavor and recreating a series of new standards based on the authentic raw taste in the production regions. Compared to industrially produced Menghai tea, the raw tea now represents a higher level of culture.

Rather than favoring either kind as uniquely authentic, these connoisseurs' standards suggest that both the raw and the aged are authentic, as long as the raw is processed in the same way as the aged might have been. Time will bridge the gap, and the raw will one day become as valuable as the aged. In this regard the renaissance of the Puer tea industry relies largely on deduction and imagination, and the originality of Yiwu and its tea becomes "imagined originality."[14]

CHAPTER 2

Tensions under the Bloom

Since he is in *jianghu,* he could do nothing but follow the law of *jianghu.*
—Paraphrased from a Chinese proverb:
Ren zai jianghu, shen bu you ji.

I arrived in Yiwu at dusk one day in early March 2007 to start my fieldwork. After four hours on the bus from Jinghong along winding mountain routes, I was tired and hungry. Accommodation was unexpectedly hard to find. Most of the local guesthouses were full, and when I finally found one with a room, even after tough bargaining, the monthly rent was twice that of a good apartment in Kunming.

While I ate dinner at a restaurant along the main street, I noticed that several houses nearby were under construction. Although it was evening, trucks and cars were still passing by, kicking up dust. As I watched the busy road, I wondered whether Yiwu's booming business was largely a result of the Quality Safety Standard, a regulation issued by the state government around 2006 to standardize the tea production process.

The next day I went to visit Mr. Zheng, an old man whom I had met one year earlier. I recalled that his house was pleasant and large, with several separate rooms, an open yard with a small table for eating and drinking tea, a lovely front door framed with orange flowers on a vine, and a long stairway leading down from the house to the street. His house, however, seemed to have disappeared. Fortunately, while looking for the house, I ran into Mr. Zheng, who recognized me and invited me in. The house still stood in its original place, but the stairs, front door, and orange flowers had all disappeared, replaced by a paved slope for the convenience of tea transport, according to Mr. Zheng. Five meters away, a pickup truck was parked under a tree. Mr. Zheng's family house was also greatly transformed. Two-thirds of the internal spaces, which used to comprise several bedrooms and one living room, were segmented into smaller units for tea processing on a production line. There were individual rooms for storing, sorting, pressing, and drying tea leaves and cakes, and a dressing and cleaning room for workers. The size of the rooms ranged from about 4 to 10 square meters. Only one-third of the original house was left for living spaces. Bedrooms were in short supply and could not accommodate guests, even immediate family members who came home to visit.

I lamented the narrowing of the Zheng family's living quarters, but Mr. Zheng said that only in this way could he meet the Quality Safety Standard without building another tea-processing house elsewhere. Another building would have cost more than ten times as much as he had invested in renovating his house.

Puer tea production in Yiwu is divided into two stages: rough processing and fine processing. Rough processing includes harvesting the tea, killing the green (stir-roasting tea leaves to suppress fermentation), and rolling and drying the leaves. The final product at this stage—loose dried tea leaves—is called *maocha*. In fine processing, *maocha* is steamed, shaped by hand in a cloth bag, pressed with a stone, dried, and packed. Before the Quality Safety Standard came into force, these procedures, except tea harvesting, were all done at local family houses. Processing areas were not routinely separated, and processing occurred under the same roof where people lived and chickens were kept.

Starting around 2004, the National Administration of Quality Supervision launched a series of standards to bring more food products into line with a stricter set of market standards. Accordingly, the Yunnan Provincial Supervision Bureau of Technology and Quality began to draft the Quality Safety Standard on tea in 2005.[1] While it was rumored that this was just a new way of collecting additional taxes from the growing Puer tea industry, formal documents declared that all the requirements were significant in assuring that tea is processed in a clean and safe environment (Zhang Shungao and Su Fanghua 2007: 207). What the standard actually aims to regulate is fine processing. It has strict requirements for processing sites, scale, and facilities. For example, it requires that processing sites be at least fifty meters away from refuse dumps, farm animals, and hospitals, at least one hundred meters away from fields where pesticides are applied, and even farther from industrial areas; that processing areas be a certain minimum size; that there be separate rooms for raw materials, auxiliary materials, and semifinished and finished products, with no other goods stored with them; and that specific machines be used for different aspects of processing (Zhang Shungao and Su Fanghua 2007: 222).

Most of these standards challenged the "traditional" way of fine processing in Yiwu in terms of scale. Fine processing had taken place in family houses since the establishment of the old commercial brands in the Qing and Republican periods. After lying dormant for almost half a century, these fine processing techniques had been revived following the Taiwanese visit. The local producers followed the example set by their ancestors, who found nothing improper about making caked tea in small family units. The new standards, which those in the industry called QS in abbreviation, however, would transform small family spaces into large-scale units, more like tea factories, leaving little living space in family homes. Most importantly, the new standards required a high level of investment. Although the Puer tea renaissance had helped locals to improve their standard of living, it was still a major challenge for most families to spend money on enlarging their tea-processing facilities.

The Quality Safety Standard was formally announced in Yiwu in February 2006, with the message that if producers did not conform by the beginning of 2007, they would no longer be allowed to produce Puer tea and their products would be banned from the market. The announcement caused considerable panic. Many people were afraid that they would not be able to continue to produce Puer tea.

Under the double pressure of the modern standard and oppositional calls to protect "tradition," a compromise emerged. It was proposed by some elites and officials of Yiwu and Yunnan and endorsed by the relevant governmental departments in Xishuangbanna. The result was the founding of the Yiwu Zheng Shan (Authentic Tea Mountain Yiwu) Limited Company in June 2006. Twenty-four family units (around one-third to one-half of the total tea-processing units in Yiwu)[2] plus one general company in Kunming participated as the stockholders. All member units were expected to try their best to transform and improve their tea-processing environment based on their original family scale, but they did not need to build a new processing area. The transformation and improvements were implemented under the guidance of the company and the QS supervision bureau. After all the family units finished their transformation and passed the government checks, the company was granted a single certification of QS approval. That is, twenty-five tea-processing units shared the same QS certification. Mr. Zheng was one of the participants in this organized unit, and as a result he didn't build a new tea factory but just renovated his family house.

I visited other local tea production units that were members of this company as well. Like Mr. Zheng, all had tried their best to transform their living spaces in order to comply with the standard and be able to continue their business. But I still could not help being shocked when I entered the old house of Mr. Li, a retired public servant. The central town of Yiwu is divided into two parts. To the south of the main street is the comparatively new area near the vegetable market. To the north is the so-called old street, where traditional houses stand (map 2.1). Mr. Zheng's house was located in the new area and was built during the 1950s in a more modern style. The transformation of his house was "all right," in Mr. Zheng's words, since the house was not old enough to warrant historic preservation. But Mr. Li's house, built in the early 1900s, was one of the most famous examples of Han architecture on the old street (fig. 2.1). It had been owned by a political leader during the 1930s and early 1940s and then used as the office of the local authorities in the late 1940s. Mr. Li, who was seventy years old and had once worked in the Yiwu government office, told me in 2006 about the history of his house. At that time, the two-level house looked old but was still well kept, with a courtyard surrounded by ventilated corridors. But by March 2007, part of the corridors were blocked, separated and marked as tea-processing rooms, and the new white cement applied to the renovated sections obviously contrasted with the worn color of the original house. From other locals, I learned that Mr. Li should not have been allowed to be make any big transformations on this house, since it was on the list of Yiwu's historic buildings. But the local government was unable to advise Mr. Li on how to maintain the house's original architectural features while complying with the new standard. Eventually Mr. Li joined the Yiwu Zheng Shan Company and renovated part of his house for tea processing. Since he had worked in the Yiwu government and was well respected, no one formally objected to his transformation of the house.

However, Mr. He, who shared the other half of the original ancient house, took an entirely different attitude and refused to join the company. The changes it had recommended, in his opinion, would neither protect

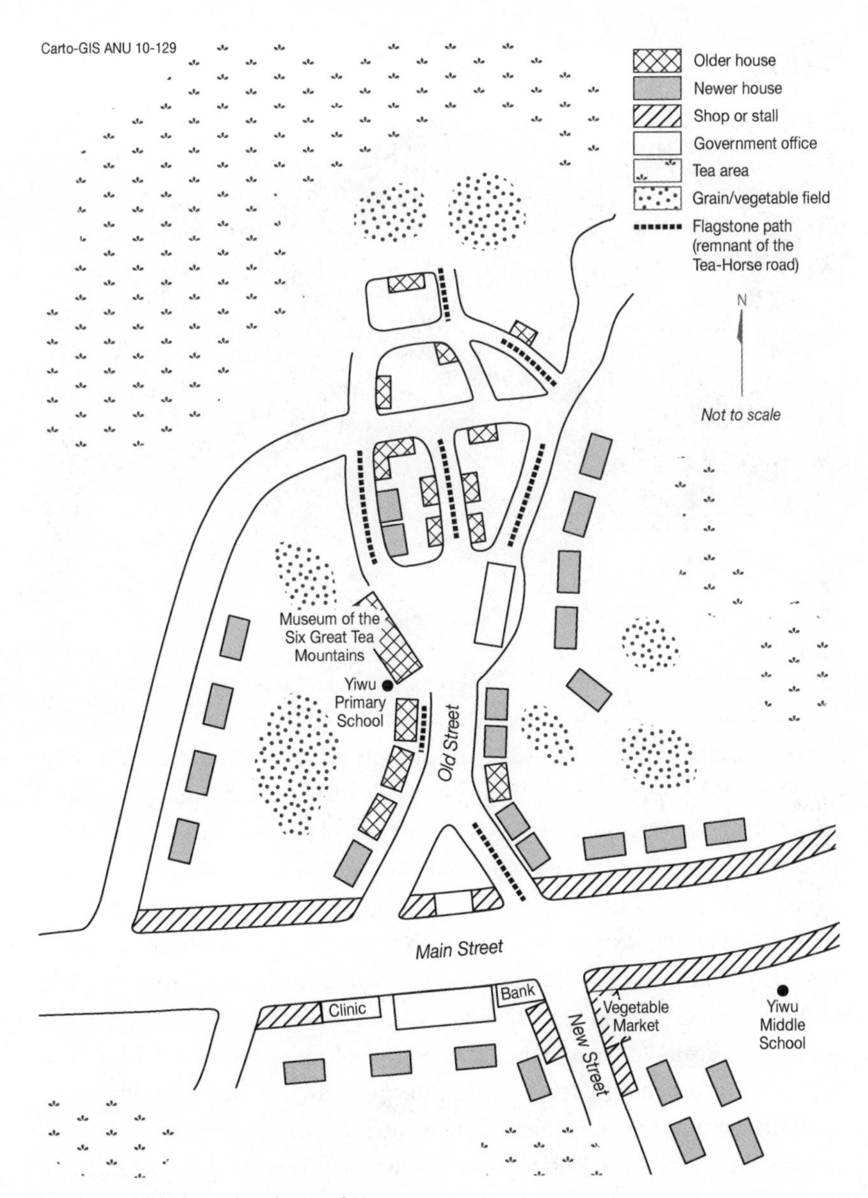

MAP 2.1 A sketch map of central Yiwu.

FIG. 2.1 Mr. Li's house is one of the "preserved heritage" houses in the old street of Yiwu (in 2007). Photo by Sun Jingfeng.

the traditional architecture well nor comply with the new standard in a strict sense. Mr. He told me that several months earlier some provincial officials had come to assess the implementation of QS in Yiwu. After visiting processing units that looked more like factories and had passed QS, they came to Mr. He's house to observe traditional production of Puer tea in a well-preserved house. Their visit had made Mr. He proud, but his application to local authorities for assistance in acquiring QS without destroying his house's historic architecture was not accepted. This angered Mr. He but he still preferred not to take the same course of action as his neighbor. His daughter was opening a tea shop in Kunming, and he had built up stable relationships with outside clients. He wanted to continue his tea business, and he had to pass the QS. He seemed worried as well as confident. His confidence, I gradually learned, came from his collaboration with a large tea company in Kunming that had promised to sponsor him to build a completely new tea factory elsewhere in Yiwu.

According to the Yiwu government, in March 2007 there were fifty units devoted to fine processing of Puer tea in Yiwu, and thirty-six of these had passed the QS in various ways.

BLOOM AT HIGH PRICE

Tea harvested in spring is regarded as the best because the plant accumulates more nutrients after a winter rest. Around March 2007, various groups involved in the tea trade converged on Yiwu. Traders, travelers, journalists, photographers, and artists came from all over the world—from Jinghong, Kunming, Beijing, Shanghai, Guangdong, Henan, Hong Kong, Taiwan, Japan, South Korea, France, and America—in a scene that seemed to echo the imagined historical glory of Yiwu, when "hundreds of thousands of people came to trade in the tea mountains" (Tan Cui [1799] 1981: 387). The guesthouse where I stayed was bustling. People came and went, but few interacted, and suspicion prevailed between visitors and local families, as well as among visitors themselves. A trader from Kunming told me after we had become friends that he suspected that I had been invited by the Zheng family to collaborate in their sales promotion. This atmosphere made me feel like the kind of traveler found in martial arts fiction at a roadside inn, where strangers of unknown origin meet, each anticipating a fight. Indeed, the Puer tea business is like the world of *jianghu,* with suspicion between people arising in a context of heightened competition. Trading authentic Puer tea in Yiwu was becoming a major challenge.

Most travelers stay in Yiwu for only a few days, but traders must stay longer. They must select the basic tea material, find a trustworthy local family to process it, negotiate a price, and finally sort out packaging and transport. Since 2007, traders must also verify that the local family they are working with has met the Quality Safety Standard or their final tea product will be hard to sell in the urban market. (In fact, a few traders still collaborate with locals without QS, but these traders must sort out their own ways of selling.)

By 2007, Wen, a trader from Kunming, had operated his Puer tea business for five years. Every year he came to Yiwu before the real business began. He spent time making inquiries, bought *maocha* from select mountains at carefully chosen times, and supervised the whole procedure of fine processing, from tea pressing to packaging. After the final products were dispatched, he left Yiwu, usually at the very end of the harvest season.

In 2007, what initially concerned Wen and the other traders was the abnormal climate: it was rather dry in Xishuangbanna that spring. As a result, the tea plants were sprouting slowly and *maocha* was selling for a higher price. The highest price the previous autumn had been around ¥130 per kilogram, and it would not be unusual for the price in the next main harvest season to increase by ¥30 or ¥50. However, in mid-March 2007 the starting price of *maocha* was more than ¥300 per kilogram, and one of the famous subvillages of Yiwu was selling it for ¥360.

Wen refrained from buying in March, hoping the tea price would decline later. He attributed the abnormally high price to two factors: the dry weather, which had resulted in a limited supply that could not match the high demand, and the "interference" of a few tourists, who had impatiently bought *maocha* before the real traders could start their work.

One day in mid-March, Zheng Da, Mr. Zheng's oldest son, came back from Jinghong with news that *maocha* in Laobanzhang—one of the tea mountains near Menghai, to the west of the Mekong—had sold for ¥800 per kilogram two weeks earlier. There was a rumor that this high price had been set by the local government and that it had suddenly jumped to ¥1,200 one day before Zheng Da came back to Yiwu. Others said that whereas in the past guests would always be served tea in Laobanzhang, they were now given a bottle of water, because tea had become prohibitively expensive.

At this time, people began to realize that the increasing price was not confined to Yiwu but was common in all the tea areas of Xishuangbanna and Yunnan. In fact, it seemed that Yiwu's price was not that high at all and could rise further to catch up with prices elsewhere. Some tea farmers and outside traders also said that the surprisingly high price of Laobanzhang *maocha* was caused by fierce competition among Guangdong traders, who liked the strong taste of Puer tea. Influence from other tea areas, therefore, was taken as a third factor contributing to the rise in price.

The increasing price of *maocha* brought worries to some and happiness to others, depending on their position. In Yiwu, almost every family was involved in the tea business, whether in rough or fine processing. The higher prices made those who were involved only in rough processing happy, since they needed only to harvest their tea trees, process the leaves into *maocha,* and then sell the *maocha* at a high price without worrying about later procedures. This boosted their enthusiasm for working in the tea fields. People joked that a tea picker coming back from the fields was "carrying a basket of money in the street" or "rolling money very hard," referring to one of the rough processing techniques wherein tea leaves are rolled by hand.

For those involved in fine processing, however, the increasing price of *maocha* brought concern. They now needed more capital to buy *maocha* and had to figure out how the final products would be sold; as prices rose, they feared consumers might not be able to afford them. Mr. Zheng was among those who was worried. The tea that he had produced during the last two years had sold well, but he still had some left over. And he had spent money on meeting the QS requirements: about ¥50,000 to renovate his house and another ¥20,000 to the Zheng Shan Company, of which he was a member. Now he would have to spend more money to obtain the unreasonably expensive *maocha*. He was waiting for the price to decline but worried that the tea material would be snapped up by others and not be available for him later.

Unlike Mr. Zheng, another local producer, Mr. Gao, was involved in both rough and fine processing. He owned some tea fields, but he still needed to collect more *maocha* from other sources for his larger-scale new factory, established in partnership with a friend. The new factory had cost them ¥600,000. After having passed the QS standard successfully, it opened in late March 2007 and was proclaimed to be the best in Yiwu. But Mr. Gao and his friend had borrowed money to build the new factory and were now in debt. Because the price of *maocha* had increased unexpectedly, they would face even more debt.

Many of these local tea producers adopted a neutral strategy: they began to buy both cheaper, lower-grade *maocha* and expensive, high-quality *maocha* so that they would have excellent as well as ordinary products to supply later. Their primary goal was to find stable clients to whom they could sell their tea. They were happy that Puer tea was becoming more popular, but they realized that the competition in the business was heating up. They could see that there were many outside traders who came to Yiwu to collect *maocha,* with whom they competed as well as collaborated.

Most outside traders' involvement in fine processing was limited to contracting with tea producers in Yiwu, and the unexpected rise in prices worried most of them. For many individual tea traders, and even some bigger tea companies, not knowing whether the higher-priced tea products would be accepted by the market, and therefore whether they'd earn enough to buy additional *maocha,* was a source of great anxiety. Only a few traders were happy about the situation, such as this one from Beijing:

> Of course I feel the price is increasing abnormally, but I am happy because it means that a large number of incapable traders will be defeated and finally thrown out of the Puer tea market if they are not able to cope with the challeng-

> ing price. And I am confident about myself. This is just a shuffling of the cards, where only the fittest will survive.

Something I heard at the end of April echoed this statement. Mr. Gao conveyed some important news: "I heard that three big companies from Kunming have come to Yiwu this spring. After deliberately raising the tea price by driving out others, they left." So besides bad weather, short-stay tourists, and price competition between tea areas, the deliberate interference by outside companies was the fourth reason for the high price. This was never verified, but it persisted as gossip. It also made the tea price increase in Yiwu more mysterious, as an invisible hand seemed to be "shuffling the cards."

What was obvious was that the price of everything in Yiwu was increasing. When I went to Yiwu's only photocopy shop, I was charged five times as much as I would be in Kunming for printing and copying. I was told that "since the tea price has been elevated, we have to follow." There was no exception for meat, vegetables, or guesthouses. In 2007, the price of pork and many foodstuffs in China rose sharply, and it was hard to determine whether the tea price followed the prices of these other commodities, or vice versa.

In addition, the number of Puer tea investors was increasing. One afternoon in early April, Wen, the trader from Kunming, came to Mr. Zheng's house for tea. Just before the tasting, another man came in and said hello. He introduced himself as Xu, a tea trader from Guangzhou, and said that he had arrived in Yiwu that morning. It was clear that he was trying to find a local family to cooperate with in the future. Mr. Zheng warmly invited him to join in the tea tasting. Unlike many traders, who resisted sharing their background and purpose at the beginning, Xu revealed that he had traded tea in many areas of Yunnan and thus could handle various requests for his clients. He described the journey to Yunnan as arduous, almost as difficult as the caravan journeys of the past. To save money, he had taken a train from Guangzhou to Kunming and arrived in Yunnan in early March. He had traveled through Baoshan, Lincang, Simao, and Menghai before finally arriving in Yiwu (maps I.1 and I.2). However, during his journey, shocked by the continually rising price, he had not collected much *maocha*. He was disappointed that he had achieved nothing in the months since he had left home. Tortured by the high price of *maocha* and unable to understand its change, he had on several occasions almost decided to turn back and give up, but he thought about his long-term business and persisted in this hard journey. He sincerely hoped to do some business in Yiwu by cooperating with a local tea family. But since Yiwu was totally new to him, he was eager to gather local information.

Moved by the story of Xu's hard journey, Mr. Zheng and Wen told him about Yiwu, and Mr. Zheng showed him the tea products his family had made. Looking at the Puer tea cake, Xu gave a deep sigh. He said that he was forced into the business by requests from his clients, since Puer tea was in heavy demand in Guangdong. He had a wholesale tea shop in Fangcun Tea Market in Guangzhou, the largest wholesale tea market in China. But he described himself as a small fish (*xiao yu*) swimming among thousands of tea traders in that market, where everyone was now involved in the Puer tea business. His favorite tea, he admitted, was actually Iron Goddess of Mercy (Tieguanyin), a famous oolong tea, which was his dominant product prior to Puer tea's prevalence. Out of politeness he continued to drink the Puer tea served by Mr. Zheng, but he showed little interest in it and seldom commented on its flavor. Comparing Iron Goddess of Mercy with Puer tea, he said the former had risen steadily in price and was a mature tea product in the market, unlike Puer tea, whose ascent was chaotic and whose prospects were questionable.

Xu did eventually conduct business with Mr. Zheng. He collected *maocha* by himself or deputized Mr. Zheng to collect it for him. The fine processing was done at Mr. Zheng's house, for which Xu paid ¥10 per kilogram. Xu traveled continually between Yiwu, Menghai, and several other tea areas, complaining about the hard journey and tough prices. Xu's case was typical and shows that even those who did not have a real interest in Puer tea and were only "small fish" had been forced to join in the sea of the Puer tea business.

I later met more people whose past work had nothing to do with tea but who had grown enthusiastic about Puer tea trade or collection. With limited tea resources and a greater number of buyers from more and more places, it was no wonder that the price of *maocha* was rising higher and higher.

Puer tea and its increasing price dominated events in Yiwu during the spring of 2007 (fig. 2.2). People talked about tea at meetings, over dinner, and even at weddings. Many local young people decided to stay at home and help their parents with the tea business rather than move to cities, where they might earn less than they did in Yiwu. At the same time, workers from

FIG. 2.2 *Maocha* being sun-dried in the main street of Yiwu. Photo by the author.

other rural areas—with diverse ethnic origins like Yi, Miao, and Han—were coming to Yiwu to enter the tea industry.

"A BATTLE OF WITS AND BRAVERY"

Rain eventually broke the drought in Yiwu in mid-April, but the price of Puer tea continued going up. By the end of April, when the tea leaves started sprouting more quickly, the price of *maocha* had reached ¥400 per kilogram. Wen decided that he would begin collecting *maocha* before the spring season was over. The task of identifying "authentic" *maocha,* which had been a challenge before, would be even harder this year as a result of the intense competition.

Among the various traders, Wen was above average. He said he ran his individual business on a small scale, but he emphasized that he had the choicest goods. And he was confident that he would not be shuffled out of the Puer tea market because he had stable customers, especially those from Taiwan who demanded high quality. These clients had preordered and prepaid for Puer tea. They trusted Wen, and they accepted that the price he charged would fluctuate in line with market conditions. But for Wen, trust also meant pressure. As his clients were critical tea drinkers, he had to select the best tea in Yiwu if he wanted to maintain a stable relationship with them.

Although the QS standard was established by the provincial government to ensure the clean production of Puer tea, it focused only on fine processing and left rough processing unregulated.[3] The latter, however, is considered essential by critical traders, like Wen, in determining the quality of *maocha* and hence that of the later Puer tea products. For example, how much the tea plant was pruned, at what temperature oxidation was stopped, and how the tea leaves were rolled all have an important effect on the final flavor. In many respects, these considerations are similar to the specialty coffee standards that arose in the United States in the 1970s as consumers began to demand handcrafted, single-origin coffee with distinctive flavors (Talbot 2004). After being heavily influenced by Taiwanese connoisseur standards, Wen and other traders further developed their own practical guidelines and conveyed them to the tea farmers. These guidelines supplemented and went beyond the state's regulations. Different traders guided the growers in different ways, which resulted in a diversity of methods of rough processing tea in Yiwu. Because the tea plants belonged to local farmers, Wen and the other traders could not always be there to supervise, but they offered guidance when they visited. As a result, the final verdict on the authenticity of *maocha* could be made only via careful judgment before buying.

Wen was perhaps the most selective trader I met in Yiwu, and he was skilled at differentiating among teas. First, he selected authentic *maocha* made from forest tea, which he processed separately from *maocha* made from terrace tea. Yiwu has many old tea trees, some hundreds of years old, scattered amid the forest. The oldest tree, which may have been cultivated initially by the indigenous people, is estimated to be about eight hundred years old. Since the early 1980s (or late 1970s, according to some locals), large terrace tea gardens have been planted. Terrace plants are spaced much more closely than forest trees and look more like bushes due to regular trimming, but the resulting tea retains the same botanical characteristics as forest tea (figs. 2.3–4). Historically, *maocha* made of terrace tea and *maocha* made of forest tea commanded the same price. However, since around 2004 (or 2002, according to some locals), forest tea has been traded for a much higher price due to the claim of outside traders that older plants produce tea superior in flavor. Forest tea is thought to be minimally contaminated by pesticides and fertilizer due to its growing in a healthy ecosystem under the forest canopy

FIG. 2.3 Forest tea. The post is to help with climbing to pick the tea leaves. Photo by the author.

among other plants. Moreover, the most aged and valuable Puer tea, like Tongqing and Songpin, was made from forest tea material before terrace tea fields were established, which also contributed to a price gap between forest tea and terrace tea. In 2006, forest tea was valued at more than ¥100 per kilogram, while terrace tea commanded around ¥30 or ¥40 for the same amount. When the former reached around ¥400 in the spring of 2007, the latter was around ¥100.

Differentiating forest tea leaves from terrace tea leaves became more challenging when the price gap widened in 2007, and it became more common for terrace tea to be passed off as forest tea. According to a rough estimate by Zhang Yi and his clients, Yiwu's annual forest tea output was less than one hundred tons, and around three hundred tons if terrace tea was included, but in the market there were 3,000 tons of purportedly "authentic Puer tea from Yiwu." Under these conditions, it is nearly impossible for the ordinary consumer to identify authentic forest tea of Yiwu. The buyer must first carefully examine the tea's appearance. Some scrupulous buyers, such as Wen, claim that they are able to determine the origin and quality of the

FIG. 2.4 Terrace tea. Photo by the author.

tea simply by sight. According to them, forest tea has larger leaves and stems, with tiny "fur" covering the back of the leaves. Others contend that terrace tea grown with fertilizer may have the same appearance, or perhaps look even better. Therefore, most careful buyers rely on flavor differentiation, infusing and tasting the tea themselves. Both good forest tea and good terrace tea have a sweet flavor followed by bitterness. But forest tea's aftertaste is said to be more intense than that of terrace tea.

A second test of "authenticity" involves selecting *maocha* that was grown within the Yiwu area. Wen declared that Yiwu tea is one of his favorite types of Puer tea because of its subtle, lingering flavor.[4] He processed *maocha* from elsewhere separately from Yiwu tea and labeled each tea by region, as his clients expected. Most tea producers and traders followed, or at least claimed to follow, this approach. Strictly speaking, they also agreed that tea material from the other nearby five tea mountains should be processed separately. But most considered these tea mountains to be "brothers" of Yiwu and tolerated the blending of *maocha* from all six mountains, including Yiwu. More critical traders, like Wen, strongly opposed blending Yiwu *maocha* with *maocha* from "nonbrother" regions, such as Jiangcheng (a county of Simao, north of Yiwu), Shangyong (another township of Mengla, south of Yiwu), and across the national border in Laos (east of Yiwu) (see map I.2).

But growers in Yiwu and the nearby tea mountains could not meet the increasing demands of local tea processors for Yiwu *maocha*. As a result, many traders secretly bought *maocha* from nonbrother regions but retraded it as "authentic" Yiwu *maocha*. This adulteration made it even more difficult to identify true Yiwu tea.

The third challenge for Wen was to identify the particular sub-tea mountain within Yiwu in which the *maocha* was grown. In many local family houses, there was a map of the Six Great Tea Mountains, showing the location of tea resources, relics, past caravan routes, and present transportation lines. The areas of Yiwu in this map were remapped in Wen's mind as distinct tea areas, characterized by trivial variations in taste determined by their distinct ecosystems. In Wen's mental map, many areas surrounding central Yiwu produced substandard tea resources. Only one western area (Gaoshan village, inhabited by Yi tea growers), one northern area (Ding village, inhabited by Yao), and one eastern area (Guafeng village, inhabited by Yao, very close to Laos) were considered adequate by him. These areas were more sparsely populated and had more undisturbed forests. According to Wen, Han tea growers often used a superior rough processing technique, but the better ecosystem areas inhabited by Yi and Yao produced superior tea material. He believed that forest tea originating from these three areas had a unique cool flavor and a longer aftertaste than tea from other sub-mountains within Yiwu.

Long before enough *maocha* was available and when the price was still unstable, Wen visited the tea-growing villages several times—talking, negotiating, and building relationships with local Yi and Yao growers. He even helped guide the growers in rough processing methods such as stir-roasting and rolling. In early March I joined Wen and several of his clients from Guangdong on a visit to the Yi village of Gaoshan. These clients had previously bought Puer tea from Wen that had been made from Gaoshan tea material, and they now wanted to see Gaoshan in person. After we visited the tall tea trees of Gaoshan, Wen took us to the house of one of his local acquaintances. There, the master, Xiao Hu, served tea, which Wen praised highly, saying that this was the *maocha* he would collect. Later, the family prepared lunch and invited everyone to join them. The guests were embarrassed to eat without paying, but Wen invited them to join on behalf of the master, saying that his guests should make themselves at home. Before leaving, Wen reminded Xiao Hu not to prune the tea trees[5] and not to mix any terrace tea with the forest tea. He said that once a sufficient amount of *maocha* had been prepared, he would return to buy it at their negotiated price.

At the end of April, when Wen decided to begin buying *maocha* in earnest, I followed him again to Gaoshan. On our way to the village, Wen told me that the price of the *maocha* would be ¥430 per kilogram, as Wen and the growers had agreed. This was a little higher than the average price of *maocha* in central Yiwu, but Wen said it was worthwhile because Gaoshan's *maocha* was better and Xiao Hu had improved his processing technique (largely as a result of his instruction, Wen hinted).

Xiao Hu was not at the family house when we arrived, but his father and brother brought out three big bags of *maocha*. Wen checked the bags one by one, carefully examining, smelling, and touching the leaves. At one point, Wen frowned, because he discovered some terrace tea mixed in to one of the bags. Xiao Hu's father did not deny this but said that only a little terrace tea had gotten in and that it was of good quality. He said that it had been mixed in by accident by his wife. Wen tried to restrain his feelings, repeating the requests that he had made on his earlier visit. Later, however, when the family demanded a new price of ¥460 per kilogram, rather than the agreed-upon ¥430, he lost his temper.

Xiao Hu's father explained that this new price was now being charged by many families in Gaoshan, but Wen exclaimed and shook his head. The previous oral agreement had been broken and it seemed that the trade would be terminated. Xiao Hu's father refused to lower his price. As I understood later, if Wen did not buy at this moment, the family could easily sell the *maocha* to another trader. The future price of Gaoshan's *maocha* was even harder to predict. And rejecting the price would mean breaking the trade relationship with Xiao Hu's family.

Xiao Hu's father made a small concession, saying that he would sell the bag mixed with terrace tea at the old price of ¥430 per kilogram, but he insisted on selling the other two bags at the higher price. Wen thought for a while, writing something in his notebook. Finally he accepted. The *maocha* was weighed, and the price was calculated. Wen and his assistant carried the three bags of tea to Wen's car, and Wen left without further conversation with Xiao Hu's family.

On the way back to Yiwu, Wen told me that something similar had happened to him in the Ding village inhabited by Yao, where he had donated many study items to the local primary school. The Yao people, Wen admitted, had been impoverished for a long time, but their situation was improving as a result of the Puer tea trade. He thought he had treated them fairly and did not understand how they could be so greedy. As for Xiao Hu's family, Wen had deposited some money with them many days ago as a promise that he would buy their *maocha*. What could he do? The process of dealing with these people, he said, was "a battle of wits and bravery."

Although the price of tea had grown incredibly high and the competition intense, Wen had hoped to succeed based on his own ability. However complicated the *jianghu* of Puer tea in Yiwu, Wen, like the Chinese knight-errant, believed that he could solve problems using his personal skill. But however hard he tried, sourcing 100 percent authentic *maocha* was an elusive ideal. As Puer tea's popularity skyrocketed, it became harder for Wen and other traders to obtain perfect tea material. Business relationships became unstable, and it was not easy to fix them once they were broken. Finding authentic Yiwu Puer tea went beyond the skill of tea differentiation and became a struggle over human relationships.

STRUGGLING FOR THE AUTHENTIC YIWU

Visitors were eager not only to collect "authentic" Yiwu Puer tea but also to see the "authentic" Yiwu. In 2005 Yiwu was declared a "Special Tourism Village" of Yunnan, as part of a program to bring more tourists to small places with specific cultural attractions. However, the unpackaged experiences often failed to live up to the packaged versions. The contrast between propaganda and experience, between imagination and reality, between the perfect past and the uneven present led to great disappointment and even anger for visitors.

One such tourist, a nostalgic woman from Guangzhou, had imagined before her journey that she would be able to sit on the flagstones of the Tea-Horse Road in Yiwu, fantasizing about caravans ringing their bells. But in reality, those flagstones had been mostly destroyed, and in their place she found construction and dust. Many old houses had been destroyed, too. It seemed to this woman that most of the historic and valuable things in Yiwu had not been preserved well at all. I discussed this with her at the newly built tea museum in Yiwu, which collected and displayed old-style implements relevant to tea processing and transport. To this woman, who had hoped to see these relics in their original places, the exhibition made them seem dead. As Marilyn Ivy writes, "The loss of the nostalgia—that is, the loss of the desire to long for what is lost because one has found the lost object—can be more unwelcome than the original loss itself" (Ivy 1995: 10).

These complaints were echoed by another trader, Sin, from Kunming, who had been to Yiwu several times in the previous three years. He thought that the image of Yiwu was deteriorating and was distressed that more modern architecture and new construction had emerged in the old area. In his opinion, the new architecture had destroyed the "harmony" of the old street and conflicted with the traditional roofs. He took a series of photos, which he called "inharmonious" (*bu hexie*) pictures of Yiwu. According to Sin and others, the inharmonious scenes were largely the result of the QS standard and the lack of a unitary development plan for Yiwu as a whole.[6]

In May 2007, a television crew from Guangdong came to Yiwu to make a documentary about Puer tea and its original place. Upon seeing the sad state of the old street, their enthusiasm was dampened, but they had to continue the program as planned, including shooting a scene about processing caked Puer tea under a traditional roof. Unfortunately, on the day of the shooting they were unable to find a family on the old street that was processing tea. Most families in the area who used to make tea in their houses had moved to one of the new tea factories, which did not fit with the requirements of the film crew. Those families who continued to process tea at home, such as Mr. He and his neighbor Mr. Li, weren't processing that day. Finally Mr. Li kindly promised to arrange a special processing if the crew really needed it. This was the crew's only choice, but even it was unsatisfactory because modern elements had been added to the house. The filming caused a small panic for both the film crew and those locals who were present, who suddenly realized that the "tradition" of Yiwu had been seriously damaged.

Not only was the dream of a nostalgic tour destroyed, but the quality of the tea products themselves came into doubt. Many visitors began to complain when they found improper or inconsistent methods of rough processing that didn't follow the "authentic" descriptions given in guidebooks or recounted to them by tea experts. This led to further hostility toward the issue of QS. As one traveler from Beijing commented, "It is ironic that the government took great effort to ensure the cleanliness of fine processing but they don't care about rough processing. I think rough processing is more important and the quality of *maocha* is the key node for the production of Puer tea as a whole."

As the price of tea ascended during the spring of 2007, visitors were informed that some significant changes were happening in the tea fields of Yiwu: pesticides were being applied, especially on the closely planted terrace tea; fertilizers were increasingly being used to speed growth; some farmers overworked the tea plants, despite the common belief that picking tea leaves too often would result in an insipid taste; and large areas of forest were being cut down or burned in order to establish more tea plantations. All these problems of ecological destruction were attributed to the unsustainably fast growth of the Puer tea business.

What bothered ordinary travelers, like tea traders, more than anything else was the crisis in trust between people. They had not been given suggestions in any guidebooks on how to manage this.

Wang, an enthusiast for Puer tea, along with two of his friends, came to Yiwu from Henan, in central China. They declared their trip a pilgrimage (*chaosheng*); in Wang's words, they had come to find out the truth about tea in the hometown of tribute Puer. I first met them at a family house where Puer tea was served. The next day, when I met them at a restaurant, they told me that they had been suspicious about the quality of the tea served the day before by the local family. Knowing that I was doing research in Yiwu, Wang said he felt it was necessary to tell me what he had learned in Yiwu. He said he was disappointed by the place, and especially by the people:

> Even at the origin of the tea plants, the hometown of tribute Puer tea, I still could not find out the truth about it. Although the trip gave me opportunities to taste some comparatively pure Yiwu tea, I still could not tell what on earth good Puer tea was. The criteria is in chaos even in the tea field; it is no better than the situation in the urban market. Most locals are hospitable and honest, as long as you don't mention the tea business. But because of the increasing value of Puer tea, the local tradition has been lost. Everyone in Yiwu is after a profit, and little clean earth remains.
>
> I had been delegated by several friends to buy some excellent Yiwu Puer tea and take it back with me to Henan. However, I now realize that I have been cheated while traveling around Yiwu these past few days. Different grades of *maocha* are secretly being blended. How could I tell my friends that I could not find authentic Puer tea in Yiwu? Henanese used to drink more Maojian [a sort of green tea mostly composed of tea buds]. It is complicated to identify, too, but

if you go to its hometown, you can discover a clear and clean version. I wonder why the same standards for Puer tea do not exist in its hometown.

Like Wang, many travelers grew suspicious of the quality of Puer tea that they had bought in Yiwu. Similar to what they had experienced in urban markets, 90 percent of Puer tea products bought in Yiwu and marked as "authentic forest tea of Yiwu" were later found to be terrace tea leaves from the same region or, even worse, blended terrace tea leaves from somewhere far away. Worst of all, bad leaves that had been poorly processed, resulting in an unpleasant taste, were still declared "authentic." Wang asked, "What sense would it make for me to buy piles of bad-quality raw Puer tea, store it, and wait for its aged value? It would waste both my money and my time."

CONCLUSION: AUTHENTICITY ANXIETY, COMMERCIALIZATION, AND CHINESE INDIVIDUALISM

As the Puer tea trade blossomed, the authentic status of Yiwu became blemished and inauthentic.[7]

In the West, the rise of modern techniques during the nineteenth-century industrial revolution, which generated mass printing, photography, and film, transcended the singular form of things and gave rise to worries about forgery (Benjamin [1936] 1999). The concern with authenticity was also seen by Western scholars as being tied to a rising modernity, the Western notion of individualism, and the emergence of private property (Trilling 1974; Handler 1986).

These views, however, apply only partially to the Chinese case, in which concern about authenticity is not necessarily linked to modernity, but is more specifically a result of increased commercialization that can be traced back centuries (Notar 2006a). Forgery was a serious concern in the thirteenth century (Song dynasty) and the late sixteenth to early seventeenth century (Ming dynasty), when counterfeits emerged in the flourishing markets and connoisseurs competed to collect antiques (Jones, Craddock, and Barker 1990; Clunas 1991; Brook 1998).[8] In the case of Puer tea, counterfeiting of the authentic had appeared in the mid-eighteenth (Zhang Hong 1998: 369) and early twentieth centuries (Colquhoun 1900: 388), which is why the producers of Yiwu stressed authenticity with special icons in their Tongqing and Songpin product descriptions. Echoing the previous periods, the recent efflorescence of the Puer tea trade has aroused anxiety about authenticity, but this is not because of mechanical reproduction, which affects only the speed of producing forgeries and relates to the preference of some consumers for handcrafted rather than mechanical products. Of greater concern is that raw Puer tea is not being made according to the new authenticity standards, nor consistent with the original aura—though many teas are also handcrafted—and hence may not be able to transform into valuable aged tea in the future.

Exhausted by Puer tea counterfeits in the cities, urban people journeyed to Yiwu, which, they imagined, would provide authenticity, original aura, and frank and transparent interpersonal relationships. However, all these imaginings about originality are contested: by the modern production regulations that contrast with tradition; by the unexpected rise in tea price and fierce competition; and, finally, by the difficulties involved in negotiating complex social relations. All these factors have seriously affected the implementation of emerging connoisseurship standards involved in finding authentic Yiwu Puer tea, and all of them reflect rising commercialization.

These concerns about authenticity can also be seen as an aspect of Chinese-style individualism.[9] Though not a dominant theme in China's history, individualism does exist and is quite evident in certain contexts,[10] as in the case of *jianghu* actors, whether in reality or in martial arts fiction. At the very least, these *jianghu* actors stand for the desire of many common Chinese to act bravely in trying situations and to find their own solutions with their own special skills. If the factors affecting the "original aura" of Puer tea production are read as the social distinctions and counterforces among *jianghu* individuals, such anxiety over authenticity appears to be rooted in conflicting desires activated in the Reform era and by rising commoditization. The desire to achieve an authentic identity and authentic lifestyle motivates the search for authentic Puer tea, while the desire to obtain wealth via tea trading and investment can lead to illegal counterfeiting and cheating. When these two kinds of desire meet, the *jianghu* of Puer tea becomes full of risk and suspicion, and the anxiety about finding authentic Puer tea goes beyond technical factors and is transformed into anxiety about the struggles of human negotiation.

Moreover, the lack of formal regulation to define authenticity and stop counterfeiting creates a vacant arena that multiple *jianghu* voices can fill, and the connoisseurship standard becomes one prominent voice arising from the popular realm. Meanwhile, however, formal regulations like the QS, although unable to solve problems efficiently, try to unify all voices and use authoritative power to enforce standards. This attempt, though not directly resisted, is deeply contested by the other actors, whether local producers who feel that an added burden is placed on them, or tourists who see QS as a modern destroyer of the original aura, or connoisseur traders whose "taste" standards diverge from those of the state. These actors' complaints, worries, and self-managed solutions (including forgery) embody multiple counterforces. As the tea economy bloomed in the mountains of Yiwu, these contested desires, multiple voices, and counterforces in turn influenced market stability, counterfeiting, and disputes over authenticity.

夏 SUMMER 热

"Killing the green" in Yiwu.
Photo by the author.

CHAPTER 3

"Yunnan: The Home of Puer Tea"

> Government has been the most powerful driving force in stoking the craze for Puer tea. . . . Government promotion of Puer tea is the most successful packaging of Chinese tea in recent years, which lies in the fact that Puer tea's history and culture have broader spaces for imagination compared with other kinds of tea.
>
> —Tang Jianguang, Huan Li, and Wang Xun 2007b: 30

Both Yiwu locals and outsiders talked frequently in March 2007 about an upcoming event that was later regarded as a key factor in the increased price of Puer tea in Yunnan that year. Another tea area to the north of Xishuangbanna called Simao, a subdistrict of Yunnan, was going to change its name to "Puer" in April.[1] "Puer" would become a confusing term, because it would mean three different things: the tea itself; the renamed subdistrict as well as its capital city, both of which had been called "Simao" in the past; and the town in Simao that had been called "Puer" but now had to yield to the new "Puer" and rename itself "Ninger" (map 3.1). (To avoid ambiguity in this book, the old terms are used unless specifically noted.)

Puer tea had long been associated with a specific place, the third in the above list. The old town of Puer had been a famous center of goods distribution and taxation in southern Yunnan since at least the early seventeenth century, when it was under the authority of the Dai state of Jinghong (Fang Guoyu 2001: 427–428; Xie Zhaozhi 2005: 3; Ma Jianxiong 2007: 563). The generic name "Puer tea" came from this town. In 1729 the Qing (1644–1912) established Puer Prefecture, whose administration included today's Simao subdistrict and eastern Xishuangbanna. The capital city of the prefecture was Puer. The General Tea Bureau (Zong Cha Dian) was established in Simao to handle matters such as taxation and tribute. The basic tea material produced in the Six Great Tea Mountains (in today's Xishuangbanna and then part of Puer Prefecture) was taken for fine processing to Puer or Simao before being sent to Beijing as tribute.

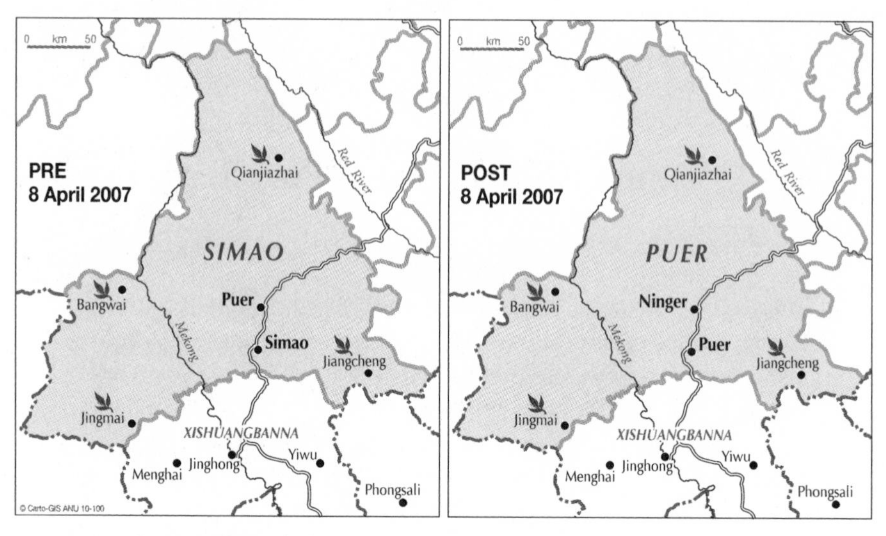

MAP 3.1 Simao and Puer, before and after renaming on 8 April 2007.

With the change of regime from the Qing to the Republic of China (1912–1949), and later to the People's Republic of China (1949–present), there have been many administrative developments in this region, which make the present boundaries of Simao (subdistrict) and Xishuangbanna quite different from those of the past. For a long time, the capital of the Simao subdistrict had been Puer, but in 1955 it was moved to Simao. In 1993 Simao (subdistrict) was upgraded from a county to a city in Yunnan. In January 2007 it was approved by the State Council of China to be renamed Puer City.[2]

The day before its name change, I arrived in Simao. The streets had been cleaned up, billboards were hung, and neon lights were turned on in honor of the coming events. The tea shops were all selling Puer tea, and local bakeries were hawking freshly made tea snacks. Small gardens in public squares used sculptures to demonstrate the tea-serving ceremony. Recognizing me as a nonlocal, the owner of the restaurant where I ate recommended that I visit the 10,000-*mu* tea gardens, a famous tourist site in Simao (fig. 3.1). The importance of tea to the city was on full display. In the local government's development plan, tea was considered the most important pillar of industry, followed by forestry, mining, and hydropower (Shen Peiping 2007).

On April 8, a large ceremony was held in Simao celebrating its name change. Ethnic performances showed the long history of the utilization by indigenous ethnic groups, such as the Bulang, Hani, and Dai, of Simao's tea

FIG. 3.1 10,000-*mu* terrace tea land in Simao. Photo by the author.

resources. A series of other activities followed in the next few days. Conferences with government officials, tea experts, and celebrities were held. Puer tea games, auctions, trade fairs, and serving ceremonies were conducted (fig. 3.2). In all these activities, the central goal was to welcome the Golden Melon Tribute Puer tea "back home."[3] This particular piece of tea was supposedly 150 years old. Weighing 2.5 kilograms, it had long been kept in the Palace Museum of Beijing as a relic of the tribute to the Qing royal family. It was unveiled at the name-change ceremony and then publicly exhibited for the next few days before being taken back to Beijing. Thousands of people went to the exhibition, curious and excited to see this royal relic originating from Yunnan. Meanwhile, in the tea trade fair, a limited supply of 999 imitation pieces, of the same shape and weight, were sold for ¥9999 each (fig. 3.3).

Many people, including me, initially felt that it was absurd for Simao to declare itself the home of Golden Melon Tribute tea, as it had been generally accepted that the Golden Melon's basic tea material originated in Yibang, one of the Six Great Tea Mountains in Xishuangbanna. Many people in Xishuangbanna were also angry about Simao's name change, since "Puer"

FIG. 3.2 Tea-serving performance at the tea trade fair in Simao. Photo by the author.

had been a common term used by many tea regions of Yunnan, and they felt uneasy that their tribute glory had been appropriated.

However, staying in Simao I found that the city had made significant preparations for this ritual, going so far as to develop a discourse legitimating the name change. This was documented in a special issue of a magazine also called *Pu-Erh,*[4] published in April 2007 by the Simao government.

The special issue presented historical details to show that the fine processing for the Golden Melon Tribute tea took place in Puer before it was sent to Beijing. This proved that, regardless of where the tea resources were from, Puer was the necessary assembly and processing place for the final tea products. Hence, Simao was of historical importance for the circulation of Puer tea as a tribute and commodity (Huang Yan and Yang Zhijian 2007: 14–17).

Second, the issue highlighted an important historical event in the local administration. In 1729 Puer Prefecture was established by the Qing, and since then the eastern Mekong of Xishuangbanna has been under Puer Prefecture's administration. Yibang and the whole Six Great Tea Mountains, now part of Xishuangbanna, were actually part of Puer Prefecture, which

FIG. 3.3 Imitation Golden Melon tea. The note on the white paper says, "No touching the precious merchandise." Photo by the author.

later became Simao. Therefore, the special issue argued that Simao was the home of the tribute tea. And the authors emphasized that the Yiwu tea family Cheshun Hao, which had been given the prized signboard inscribed with "Tribute to the Emperor" (*Rui gong tian chao*), was within the administration of Puer Prefecture at that time. Thus, although all the tribute tea did have a specific small home, Yiwu or the Six Great Tea Mountains, they were more importantly part of a larger home: old Puer Prefecture, the past Simao, and the renamed Puer City (Huang Yan and Yang Zhijian 2007: 14–17).

Third, even before the name change people had begun to believe that it was untrue that "Puer tea was never grown in Puer" but was just named after the place.[5] Proof was found to show that, in fact, Puer and Simao both used to plant and produce Puer tea (Huang Guishu 2002; Shen Peiping 2008). According to local tea experts including Zhou Deguang, whom I interviewed, there is a whole chain of tea trees located in Simao (see map. 3.1) that prove that Simao is one of the world's origins of tea: the 2,700-year-old wild tea tree in Qianjiazhai; the 1,700-year-old transitional tree in Bangwai; and the domesticated 1,000-year-old tree in Jingmai (see also Huang Yan and Yang Zhijian 2007: 17).

In addition, Simao's status was elevated in the magazine by being linked to the tea consumption of the imperial household. Historical records were produced to show that Puer tea was cherished by the Qing imperial family, praised by Emperor Qianlong in a poem, and given by the emperor to George III in 1792. Being intimate with royalty had always been considered by the Chinese to be a supreme glory, and when the royally cherished substance was returned to its origin, the origin was also glorified. At this point, the Golden Melon's homecoming was, in a way, analogous to the imperial concubine's visit home during the feudal period of China, as described in the famous novel *Dream of Red Mansions*.[6]

Moreover, the royally cherished consumption of the Golden Melon was reproduced in contemporary tastings. An account of drinking the Golden Melon written by a professor at Yunnan Agricultural University was included in the following issue of the same magazine to prove that this 150-year-old tea still tasted good, with "aged appeal" (*chen yun*) (Shao Wanfang 2007). In comparison, tribute Dragon Well green tea from east China of the same age had become worthless.[7] Once again the profile of the original place stood out, and the lasting value of the aged tea encouraged people to buy and store more Puer tea originating from Simao or Yunnan.

These testimonials made the welcoming of the Golden Melon tea an important ritual, which justified the name change from Simao to Puer and supported the declaration that Puer née Simao was the authentic hometown of Puer tea (Zheng Yongjun 2007: 9).

In response, the Xishuangbanna government commented sarcastically that Simao's name change was "good." An official working in the Xishuangbanna government said that as a result of Simao's propaganda, Xishuangbanna would also benefit, because both were important production areas. He mentioned this to me in early March 2007, when I participated in a visit to one of the tea mountains in Menghai, Xishuangbanna, where a ceremony of "Protecting the Old Tea Trees" was launched by the Xishuangbanna government. When we arrived at the central part of the tea mountain and saw many tall tea trees flourishing, this official said, "Now you can understand where the hometown of Puer tea should be; whether or not it is called Puer, it is always our Xishuangbanna, whose tea price is the highest in Yunnan every year."

On the Dai New Year[8] in mid-April 2007, shortly after Simao's name change, Xishuangbanna launched a ritual homecoming for another tribute tea, a brick of Puer originating in Yiwu. At the same time a major forum on Puer tea was held in Jinghong, its capital city.

Various maps showing alternative ideas about the home of Puer tea were published in advertisements, on tea packages, and in teahouses. Some of these maps positioned Puer as the origin and dissemination center of Puer tea. While in Jinghong, I saw another imaginative map in a teahouse that its owner, a local tea trader, had created. With an exaggerated scale and

MAP 3.2 The Tea-Horse Road, as depicted in a map in a Jinghong teahouse (English labels added).

intentionally selected details, Xishuangbanna takes up most of the tea territory (map 3.2).

The home of Puer tea was mapped by different people's imaginations, existing in their minds but being developed at a deeper level—on foot. In May 2005, a reenactment of past caravans set out from Simao toward Beijing (Baidu 2006). Over five months, 120 mules and horses with 68 drivers in ethnic costumes walked through six provinces, covering 4,000 kilometers before finally arriving in Beijing. The Puer tea carried by the caravan was auctioned there at a surprising price. Seven pieces, each weighing 2.5 kilograms, were sold for ¥28,000 to ¥120,000; one stack (seven caked pieces, each 357 grams), purchased from the caravan and donated back by the famous actor Zhang Guoli, was auctioned at ¥1.6 million.[9] In the following year, another new caravan set out from Yiwu toward Beijing. In Yiwu there is a well-known site where the past caravans assembled to start their journey. Since the start of the new caravan, this site has become an important part of the local tea tour. In various ways, home was represented with royal approval and the new proof of this royal linkage was imagined in rituals and maps and practiced in the mind and on foot.

This competition over Puer tea's origin echoed an earlier debate, which began in the early nineteenth century and lasted over one hundred years, about whether China or India (Assam, in particular) was the birthplace of tea more generally (Baildon 1877; Ukers 1935; Zhu Zizhen 1996). Most Chinese scholars insisted that tea had originated in China, using the evidence that older and wilder tea trees are found in southwest China; these scholars considered the argument in favor of India to be an attempt by the British to boost its colony's tea industry (Chen Chuan 1984; Chen Xingtan 1994). A later debate focused on which part of China tea is indigenous to. Eventually, a general consensus was reached that the tea varieties in other parts of China are mainly of small-leaf variety (*Camellia sinensis sinensis*), which originated, evolved, and were disseminated from the large-leaf variety (*Camellia sinensis assamica*) of Yunnan (Chen Chuan 1984; Chen Xingtan 1994; Evans 1992). Decades later, subregions within Yunnan continued to argue about which was the authentic home of Puer tea.

From the side of both Simao and Xishuangbanna, this debate about the origin of Puer tea seemed significant. It would directly determine which subregion was better known to the world, and in turn which would attract more traders, investors, and tourists. Both sides stressed the importance of "respecting history," implying that there existed only one authentic version of the narrative.

But from the provincial point of view, this battle seemed unnecessary. Some people thought that the success of either side would bring benefits to Yunnan, while others thought it was an unnecessary civil war because Simao and Xishuangbanna are both part of Yunnan. Actually, many people would not venture to judge which subdistrict was the true home of Puer tea; rather, they said that Yunnan, including all the subregions along the Mekong, is the home of Puer tea. When I asked a tea expert from the Menghai Tea Association his opinion about Simao's name change, he laughed and recalled a story: In one of his articles, he had identified the production area of Puer tea as Xishuangbanna, Simao, Lincang, and Dali. This, however, offended another two other areas, Baoshan and Dehong in west Yunnan, which have tea resources but were not included in his article (See map I.1). He eventually revised the article and added these two areas, and Puer tea was given a broader scope. This tea expert's action reflects the view of many people who consider Yunnan itself to be the heartland of Puer tea. Following this logic, the 2008 national standard assuring Puer tea's geographical origins put its production scope even more broadly; eleven subareas of Yunnan were all authorized as Puer tea production regions, including not only the regions along the Mekong in the south but also areas in western, eastern, and central Yunnan (NBQIQC 2008).[10]

The Puer tea battle between subdistricts of Yunnan continues to this day, with new historical "discoveries" or botanical proof being regularly unearthed. Nevertheless, there is now a broad acceptance that Yunnan is the common home for all Puer tea. However, opinions diverged again when the development of Puer tea broke through the borders of Yunnan and challenged the position of other Chinese tea.

REDEFINING PUER TEA

When Puer tea became more prevalent, its exact definition became a matter of controversy. This controversy was launched by the tea scholars of Yunnan and reinforced by the provincial government.

In the past, Puer tea had been categorized as a fully fermented dark tea. Starting around the turn of the twenty-first century, some tea experts in Yunnan began to formally argue that Puer tea is not dark tea and should be independent from the other six categories of Chinese tea (Su Fanghua 2002: 49–51; Mu Jihong 2004: 5; Zou Jiaju 2004: 9–10). The key reason, they explained, is that Puer tea is fermented using a different procedure from that used for other dark teas (see table I.1). Other kinds of dark tea, such as the one produced in Hunan province, are postfermented soon after rolling, when the tea leaves are still moist; Puer tea, by contrast, is not postfermented until after the tea leaves have dried. Furthermore, the Yunnanese experts pointed out that other kinds of dark tea are made from small- or medium-size tea leaves, but Puer tea must be made from big-leaf tea.

This updated approach to tea definition was considered an important way of exempting Puer tea from the chaotic situation in the market. As Zou Jiaju, the vice head of the Yunnan Tea Association, stated:

> Among Chinese tea, there has never been a tea like Puer, whose image is in such chaos. Many incentives are about money. Some people impute the emergence of "fake Puer tea" to the social transformation in modernity. Besides this aspect, I suppose it is also because Puer tea has lacked an accurate definition among Chinese tea. Before its recent prevalence, it wasn't really cared about, and in contempt it was wrongly categorized as dark tea. Perceptual mistakes

> cause chaos in actual trade. Other types of tea have clear identities and don't run into so many troubles. Puer tea alone is homeless and has to live under another's roof. The lack of its development is due to its unclear identity and inaccurate definition. (Zou Jiaju 2005: 135)

According to these suggestions, Chinese tea should be categorized into seven types, with Puer tea as the seventh. Because this argument was not only concerned with Puer tea's reclassification but also meant rewriting the whole system of categorizing Chinese tea, it was attacked by tea experts from other provinces.

Since the establishment of the People's Republic of China in 1949, official government statements about Puer tea have been updated several times in response to market conditions. Before the increase in demand for Puer tea, there were regulations about quality standards, rather than definitions. The first quality standard was enacted by the national government in about 1955. The second was set by Yunnan in 1979 to respond to the emergence of the artificial fermentation technique that was invented in 1973 (Xu Yahe 2006: 129). The first official attempt at a definition was issued in 2003 by the Standard Counting and Measuring Bureau of Yunnan. It defined Puer tea as follows:

> Puer tea is made of large-leaf tea leaves that have been dried in the sun. The tea leaves should be produced within a certain area of Yunnan. The final product is loose or pressed tea via postfermentation. Its appearance is brown-red; its tea brew is bright and dark red; it has an "aged aroma"; it tastes mellow, with sweetness following bitterness; and after infusing, the tea leaves are brownish red. (Xu Yahe 2006: 133)

In contrast to the academic mode of classification based on production technique, the government's regulation was more concerned with the origin of the tea. It was characterized by a strong geographical sense. This became more prominent in the next updated version, enacted by the Yunnan Provincial Supervision Bureau of Technology and Quality in 2006:

> Puer tea is the geographically marked product of Yunnan. It is made from large-leaf tea leaves that have been dried in the sun. The tea is produced in tea areas of Yunnan whose conditions are suitable for producing Puer tea. After a certain special production process, the tea develops its unique characteristics. Puer tea is categorized as Puer *sheng cha* [raw Puer tea and aged raw Puer tea] and

Puer *shu cha* [artificially fermented Puer tea]. (Zhang Shungao and Su Fanghua 2007: 313)

Both definitions, from 2003 and 2006, emphasized "large-leaf tea" (*Camellia sinensis assamica*), the tea category that is more distinctively Yunnanese than the "small-leaf tea" (*Camellia sinensis sinensis*) prevalent in other tea areas of China. Although the 2006 definition supposedly drew on academic insights, it actually contradicted the academic approach. According to the academic tea experts, whether Puer tea is categorized as a dark tea or placed in an independent category, the definition is based on the production technique (Chen Chuan 1984: 238). That is, a certain type of production process shapes a certain kind of tea. Following this rule, a product can be Puer tea as long as it is processed in the same way as Puer tea, regardless of the origin of the tea material.[11]

When Puer tea became profitable, however, the academic definition became too broad, according to the authorities in Yunnan. They worried that the fame of Yunnan's Puer tea was being illegally exploited. Large amounts of low-cost tea, mostly small-leaf tea from Sichuan, Guangxi, or Guangdong, were being carried into Yunnan or even processed locally to make Puer tea. All these products were branded as authentic Yunnan Puer tea, even though they were made from small tea leaves, a very different category from the large tea leaves of Yunnan.

Why must Puer tea be made of large-leaf tea? Zou Jiaju, the vice head of the Yunnan Tea Association, makes an analogy with red wine to illustrate. To him, both good red wine and good Puer tea have a unique mellow flavor, which is developed from the astringent element. The more astringent the grape, the mellower the wine made from it; if the grape tastes sweet, it is no good for making wine. The same applies to Puer tea. Tea made from small leaves is not astringent enough to develop a mellow flavor. Only large-leaf tea has stood the test of time and proven its vitality when aged, whether via natural or artificial fermentation (Zou Jiaju 2004: 98).

Yunnan is acknowledged as the home of large-leaf tea, but it's not the only area where this kind of tea is grown. Large-leaf tea material is also available in Guangxi, Hunan, and Hainan, though these regions cannot rival Yunnan in terms of quantity. When the availability of large-leaf tea in other areas was mentioned, supporters of Yunnan argued that the large-leaf category for Puer tea is shaped by the unique natural conditions of Yunnan that distinguish it from other tea areas. In a series of articles about Puer tea's value, microbiologist Chen Jie places great emphasis on "geographical value." To him, this criterion is an efficient way to distinguish authentic Puer tea:

> Although those fake Puer teas made of non-Yunnanese tea material greatly resemble Yunnan's, they remain at an obvious disadvantage: they can't be stored too long. The quality of Yunnan's Puer tea improves as time goes by, but the fake Puer tea made of tea leaves from outside Yunnan will only worsen. (Chen Jie 2009)

According to these explanations and arguments, the only authentic Puer tea was that which would stand the test of long-term storage, and by this logic only the large-leaf category of tea leaves, grown in Yunnan, could succeed.

But one doubt remained. What about the regions bordering Yunnan that share the same tea category as well as similar natural conditions? In Yiwu I have often observed that loose tea from Phongsali in Laos easily crossed the border to Yiwu and became part of the raw material of "Yiwu's authentic Puer tea." This tea is of the large-leaf category and shares the Mekong as mother river with the other main tea areas of Yunnan. When I raised this example to question the geographical classification of tea, the answer, from the perspective of the geographical supporters, was still "no." Tea experts admitted that tea from Laos, Vietnam, or Burma shares many characteristics with tea from Yunnan, and that, using correct production techniques, it could be made into Puer tea that would be hard to distinguish. But Xu Yahe, an expert from Kunming, emphasized that Puer tea does not refer only to a certain kind of tea with a certain production technique, nor is it based simply on a biological categorization; rather, it is a "historical geographical substance." He reminded me that Puer tea is named after a place, that Puer Prefecture was established by the Qing during the eighteenth century, and that all the tea gathered there must be taxed before being traded to other places. This, he argued, was a unique history belonging to Yunnan rather than to Laos or any other neighboring countries.

Xu's argument seemed to come back to history as the key issue. For many Yunnanese, the definition of Puer tea exists not only within the sphere of natural scientific knowledge. When natural science encounters embarrassment and cannot properly identify Puer tea with the representation of

Yunnan, history—another useful tool from the sphere of social scientific knowledge—lends a hand.

TEA CARAVANS

While Simao and Xishuangbanna debated the origin of Puer tea, Yunnan launched a series of events promoting its tea as a provincial representative in outside markets. Just as Simao used the welcoming back of tribute tea as the key ritual in its name change, Yunnan used two important propaganda techniques to present itself. One was to redefine the position of Puer tea among other Chinese teas. The other was to promote the Ancient Tea-Horse Road (Cha Ma Gu Dao). The notion that older tea tastes better was used to link these two promotional techniques.

In 1989, several Yunnanese scholars conducted a ninety-day investigation, mainly on foot, around the triangle of Yunnan, Tibet, and Sichuan. They discovered that in Southwest China there had been two main transit lines to Tibet, one from Sichuan and the other from southern Yunnan. They argued that tea was the most important commodity carried along these routes by caravan, because Tibetans drink tea to help them digest greasy food, and thus they proposed that the route be named the Ancient Tea-Horse Road (Mu Jihong 1992). Mu Jihong (2003; 2004), who led the team, subsequently extended the concept of the Ancient Tea-Horse Road to include the various caravan routes connecting Yunnan and nearby Southeast Asian areas. For example, starting from Yiwu, there used to be small branch routes to Laos, Vietnam, and Thailand, all with tea as their main transport good.[12]

The investigation of the caravan trade's history worked as a prelude to Puer tea's rise in popularity. It shed light on the geographic, economic, and political role of Yunnan as an important middle ground between China and Southeast Asia since the fourth century B.C.E. (Yang Bin 2006). As Mu summarizes:

> The Ancient Tea-Horse Road is the road that disseminated civilization, and also the channel through which commodities were exchanged; along this passage, China and the outside world communicated, ethnic groups migrated, Buddhism spread eastward, and tours and expeditions were launched. We could say that it is the ancient road with the highest topography and most complicated configuration in the world. (Mu Jihong 2004: 26)

Since the turn of the twenty-first century, numerous books, magazines, newspaper columns, web pages, and audio and video series about the Ancient Tea-Horse Road have flourished. Most were produced by local Yunnanese, who emphasized that along this passage were multiple natural landscapes, from the rainforest of Xishuangbanna in the south to the high snowy mountains in the north. There were also lively towns along the routes—Puer, Dali, Lijiang—which acted as distribution places for goods carried by the caravans. Finally, multiple ethnic groups—such as Dai and Hani in Xishuangbanna and Simao, Bai in Dali, Naxi in Lijiang, and Tibetans in Diqing—resided along the routes, each of which had a unique ethnic culture and utilized Puer tea in its own ways. In these narratives, the Ancient Tea-Horse Road embodied Yunnan's rich nature and culture and proved its significant role in the tea trade (EBCMGD 2003).

Yunnan had long been shadowed by the ill repute of drug smuggling, especially in reference to the areas bordering Burma. It had also been said to be a "backward" region with undeveloped production, transportation, and economics. In jokes, Yunnan was the place that people would go to only if they had been exiled by the emperor. The Yunnanese themselves said that Yunnanese things were "earthy" (*tu*), rustic, or backward. Or, they exclaimed that Yunnan had excellent products, such as tobacco, tea, and tour resources, but that its people were too simpleminded to package these products in eye-catching ways.

Today, when I tell people that I am from Yunnan, their eyes grow wide and they admiringly say how they wish to visit Yunnan. The Yunnanese, too, gradually realized the value of being "earthy." This transformed image developed as a result of the interplay between Yunnan and the outside as China's economic development was accelerating. After getting used to urban development, consumers began to seek something more exotic and rural, and the "earthy" element became part of a new fashionable consumption trend (Hillman 2003). The masters of the earth began to develop their self-presentation. As Puer tea blossomed, Yunnan successfully presented a series of cultural activities, the so-called Yunnan phenomenon (Yunnan *xianxiang*) (Zhu Sikun and Li Yin 2006).

In 2003 a successful stage play entitled *Dynamic Yunnan* (Yunnan yingxiang) featured Yang Liping, a famous Yunnanese peacock dancer, in the starring role. The play combined elements representing both the "primitive" ethnic folk of Yunnan and popular arts, and it became famous as it toured all around China (Fotoe 2006). *Yingxiang* (映像) literally means "image," and in Chinese its pronunciation is very similar to the word for "influence" or "impact" (影響). Through this performance, Yunnan made a dramatic entrance onto the national stage and exerted its new impact.

In December 2001, the central government gave Yunnan permission to change the name of one county in Diqing Tibetan Autonomous Prefecture from Zhongdian to Shangri-la. Yunnan had successfully forestalled Sichuan and Tibet's use of the same name. The name "Shangri-la" was initially used by James Hilton in his novel *Lost Horizon* (1939) to depict a mythical Tibetan paradise with a beautiful landscape and religious harmony. Its exact location has long been debated, but it was imagined to be in southwestern China or the nearby Himalayas. After conducting research and exploration, Yunnan announced in 1997 that Shangri-la was located in Yunnan, and the name change was approved in 2001. As a result, tourism to the prefecture and to Yunnan in general boomed (Hillman 2003). Sichuan, another area that had hoped to claim the name Shangri-la, lagged one step behind Yunnan in achieving this validity (fig. 3.4).

At the same time, more film crews came to work in Yunnan, describing it as a haven for filmmaking.[13] This praise was welcomed and encouraged by the provincial government. In 2006, ten feature films were made in Yunnan, prompting the government to remark that "the film industry is becoming one of the bright spots of Yunnan" (YRTN 2005). Among these works, some were directly relevant to the Ancient Tea-Horse Road or Puer tea. In 2004, *Delamu—Tea-Horse Road Series,* a documentary by Tian Zhuangzhuang, a fifth-generation Chinese director, was shown in cinemas; in 2005, a television series called *Ancient Tea-Horse Road* was screened by China Central Television. Both of them helped popularize the Ancient Tea-Horse Road, and many tourists came to Yunnan as a result.

Apart from new publications, plays, tours, and films, Puer tea—the old commodity of the Ancient Tea-Horse Road—became a newly fashionable drink representing Yunnan. On the one hand, narratives about the Ancient Tea-Horse Road declared that Yunnan was the original place of tea and that the demand for and transport of Puer tea had brought the Ancient Tea-Horse Road into being (Mu Jihong 1992, 2004).[14] On the other hand, Yunnanese tea experts argued that the secret of Puer tea's improvement over time was accidentally discovered when natural fermentation, shaped by sunshine and rainfall, occurred along the long, hard caravan journey

FIG. 3.4 The recently renamed Shangri-la, with the holy mountain Khabadkarpo. Photo by the author.

(Su Fanghua 2002: 50; Mu Jihong 2004: 92; Zhou Hongjie 2004: 8). With increasing awareness of the rising value of aged Puer tea in Hong Kong and Taiwan, going to Yunnan and traveling along the Ancient Tea-Horse Road became trendy in the early part of the twenty-first century.

At this time Yunnan was welcoming more visitors and also promoting more caravan expeditions and associated activities. In addition to the two promotional caravans mentioned above, another was organized in 2005 along the route toward Tibet, retracing the so-called classic Ancient Tea-Horse Road (*Yunnan Daily* 2006a). A fourth expedition, called "International Cultural Travel on the Ancient Tea-Horse Road," started from Xishuangbanna in October 2006. It was destined for Nepal, where a Yunnanese photographic exhibition was staged along with tea-serving performances (Xinjing 2006). In February 2006 an initiative was launched to store Puer tea on thirty-three famous mountains all around China (Wu Qiong 2006: 131). In July 2006 Puer tea was taken on board the *Götheborg*—a rebuilt Swedish East India Company sailing vessel—in Guangzhou. It was said that this was to link the Ancient Tea-Horse Road, represented by Puer tea, and the Maritime Silk Road between Asia and Europe, represented by the *Götheborg* (*Yunnan Puer Cha* 2006: 136–137; *Yunnan Daily* 2006b).[15]

Since the early 1990s, the Yunnan provincial government has aimed to develop Yunnan as "a province with powerful green economics," "a province with rich ethnic culture," and "a province with rich tour culture." The Puer tea industry accorded with all three aims. Being packaged with the Ancient Tea-Horse Road, Puer tea shaped the presentation of Yunnan as an important channel connecting inland China with the wider world in both the past and the present. Furthermore, Puer tea and the Ancient Tea-Horse Road were part of a provincial promotional package that involved other factors, such as new tours, plays, and films. These factors were closely linked and influenced one another. Together they were used by the provincial government to supplant the old negative impression and rebuild a new healthy image for Yunnan.

"GREAT VALUES": CHANGING CONSUMPTION AND PRODUCTION

Puer tea's improvement with age is said to be its distinguishing feature. From this, several key values were drawn out by traders, connoisseurs, consumers, researchers, mass media, and the government.

The first remarkable feature was flavor. A popular saying claimed that once you came to love Puer, you would drink no other kind of tea. I did not believe this saying until I met many people who had had this experience. Many of them told me that they used to love Iron Goddess of Mercy, an oolong tea with an enticing aroma. They admitted that they had found the flavor of Puer unpleasant when they first encountered it—either "too stimulating" (referring to raw tea) or "too earthy" or "moldy" (referring to artificially fermented tea). But as they drank more they grew to appreciate the unique flavor. The word widely used to describe the ideal flavor of raw Puer tea was *huigan,* a lingering sweetness after an initial bitterness or astringency, which was said to be much better than that from Iron Goddess of Mercy and other types of tea. Good artificially fermented Puer tea was popularly described as "warm" (*nuan*) or "smooth" (*hua*). More key words that were applied to aged raw Puer tea could be found in the book *Puer Tea,* the so-called Puer tea bible, by Taiwanese tea expert Deng Shihai. In this book, Deng (2004: 37–61) compared mature Puer tea's fragrance to that of

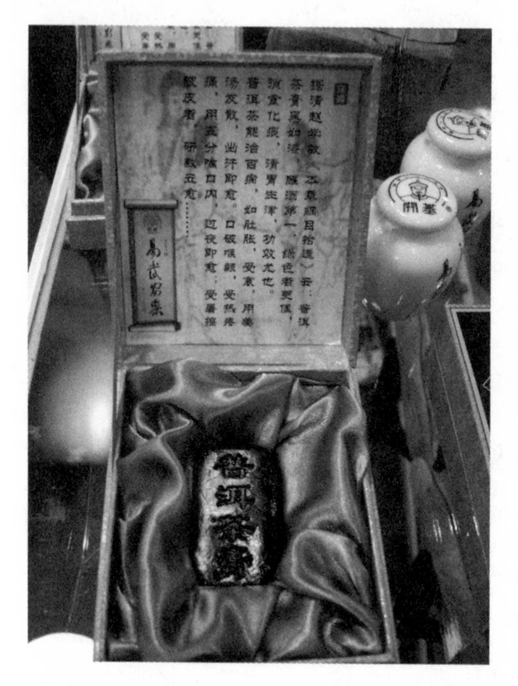

FIG. 3.5 Puer tea paste sold at the 2007 Guangzhou tea trade fair. The text on the package quotes an eighteenth-century writer (Qing dynasty) saying that Puer tea paste may help with sobering up after drinking alcohol, as well as digesting food, eliminating phlegm, and enriching saliva. Photo by the author.

orchid, camphor, and lotus and used many special terms to describe its flavor; most extraordinary were descriptions like "flavorless flavor" (*wu wei zhi wei*), "the bubbling-up of spring water from the bottom of the tongue" (*she di ming quan*), and "supplementing the vital breath" (*bu qi*). These epithets were later widely cited, and it was said that if one didn't like Puer tea, it must be because one had not yet tasted good aged Puer tea.

The second remarkable feature of Puer tea was its health value, something that almost every Puer tea drinker mentions. In interviews, several people used medical reports to show the positive effect of Puer tea. After drinking Puer tea for an extended period of time, they found that their blood pressure, cholesterol, or blood sugar was lowered. This has also been found in scientific experiments. Tea has long been acknowledged to have multiple medicinal effects, and the experiments on Puer tea tried to show that it had notable effects on losing weight, improving digestion, warming the stomach, reducing fever, lowering high blood pressure, and protecting against diseases such as cancer, constipation, coronary disease, and the hardening of the arteries (Chi Zongxian 2005; Liu Qinjin 2005; Shi Kunmu

FIG. 3.6 Pressed Puer tea sold at the 2007 Simao tea trade fair. It is in the shape of a traditional Chinese gold ingot, representing stored or accumulated wealth. Photo by the author.

2005; Zhou Hongjie 2007). Besides this published scientific research, Puer tea's health value was continually discussed in teahouses. For instance, in Simao teahouses I was told that a local tea expert had found Puer tea helpful in treating people with HIV/AIDS. I also heard that Puer tea was good for gout and altitude sickness. All sorts of information was promulgated to convince consumers that drinking Puer tea enhanced longevity (fig. 3.5).

The third widely promoted feature of Puer tea was its cultural value. This was reflected in the provincial effort to associate Puer tea with the Ancient Tea-Horse Road. In addition, Puer tea was linked to broader levels of culture: its glorifying story of being used as tribute to the emperor was promoted; it was praised for uniquely maintaining the pressed shape of traditional Chinese tea dating from the Tang and Song dynasties (seventh to thirteenth centuries); it was affiliated with Chinese traditional religion and praised as exemplifying the spirits of Daoism and Ch'an Buddhism because of its longevity; and it was used as metaphor for a way of life in which one must take time to become mature and should pursue a slow-paced and peaceful lifestyle to counteract the high speed of modernity. I once heard from an informant from Beijing that those who drank Puer tea were considered to have high "quality" (*suzhi*).[16] And generally, drinking Puer tea is said to represent a new fashion and lifestyle (Li Yan and Yang Zejun 2004).

The fourth key feature of Puer tea was its financial value. Puer tea has come to be praised as "a drinkable antique." Around 2003, many traders started doing more business in Puer tea than in green tea or oolong. It was said that green or oolong tea would lose its value if not sold in a timely manner, no matter how expensive it had originally been, but Puer tea was the opposite: the older it was, the more valuable it became (fig. 3.6). The amazing prices of Puer tea sold at auction had become widely known, and the saying that if you didn't store Puer tea today, you would it regret tomorrow spread rapidly. As a result, more and more people engaged in buying and storing Puer tea.

PUER TEA AUCTIONS

- In November 2002, at the Tea Evaluation of the Guangzhou International Trade Fair, 100 grams of three-year-old Puer tea was auctioned for ¥168,000. This broke the auction record for Tieguanyin, ¥120,000 in 2001.
- During the Chinese New Year in 2004, three grams of Puer tea stored in the Palace Museum and later collected by Lu Xun, a famous writer, were auctioned for ¥12,000. That is ¥4,000 per gram, thirty-two times the price of gold at that time.
- In October 2005, after the new caravan from Simao reached Beijing, seven pieces of Puer tea (357 grams per piece) were auctioned for ¥1.6 million.
- In September 2006, at the first Yunnan International Tea Trade Fair, one hundred grams of loose Puer tea was auctioned for ¥220,000.
- In May 2007, a piece of new Puer tea (about 400 grams) was auctioned for ¥400,000, which was so far the highest auction record for new Puer tea. (CCTV 2008)

For the reasons described above, Puer tea grew monumentally popular during the first years of the twenty-first century. This changed the pattern of tea consumption and production in Yunnan as well as in some other tea areas of China. Each of China's numerous kinds of tea has tended to establish a niche according to consumer demands, natural conditions, and food culture. Jasmine tea is more popular in Beijing and northern China; eastern China, including Jiangsu and Zhejiang (near Shanghai), has a long tradition of producing green tea, including the famous Dragon Well tea; and southeastern China, including Fujian and part of Guangdong, consumes oolong tea with great passion. On the national level, green tea has long had the highest levels of consumption in China. And before Puer tea's prevalence, oolong tea was the most popular among the middle and upper classes in urban areas such as Beijing, Shanghai, and Guangzhou. But gradually the taste, health, wealth, and cultural values of Puer tea began to exceed those of other kinds of tea. Following its revival in the mid-1990s, it took only a short time for Puer tea to become prominent in China. As a result of its rapid increase in popularity, it acquired the nickname "crazy Puer tea" (*fengkuang de Puer cha*) (CCTV 2007b).

In the Fangcun Tea Market in Guangzhou, the biggest tea distribution center in China, more and more dealers turned to Puer tea. According to statistics compiled by the Guangdong Tea Culture Improvement Association, in 2006, 99 percent of tea shops there sold Puer tea, and some who used to specialize in oolong tea also made Puer tea an important sideline. Tea transactions that year totaled ¥67 billion, with Puer tea making up one-third (*Puer Tea Weekly* 2007d).

In northern China, the new caravans made Puer tea fashionable virtually overnight. A tea trader from Beijing, whom I met in Yiwu, recalled the scene in 2005 when the first caravan arrived in Maliandao, the biggest tea market in Beijing and all of northern China:

> All the people in Maliandao were startled by the arrival of such a fantastic caravan, which they had only read about in books or seen in pictures. And what shocked them most was that a piece of Puer tea carried by the caravan could be auctioned at such a surprising price. Since then, the tea sellers in the markets began Puer tea business without exception.

In a speech, one of Simao's leaders linked the caravan to Beijing and the welcoming of Golden Melon tea "back home" (*Pu-Erh* 2007a: 5), noting that both presented an attractive image of Puer tea and its homeland, Yunnan. Jiao Jialiang, the manager of Long Run, one of the biggest Puer tea companies of Yunnan, called for "recovering the calling card of Yunnan that was lost in an alien land" (Ma Yihua 2006), referring to Hong Kong, Taiwan, Guangdong, and areas in Southeast Asia where people appreciated Puer tea much more than the Yunnanese did. To Jiao, Yunnan had always

FIG. 3.7 Decorative Puer tea sold at the 2007 Simao tea trade fair. The melon-shaped cakes, stacked in nine layers, symbolize good luck and superior status; on the surface of the round cakes are molded representations of the twelve symbolic animals of the Chinese zodiac. Photo by the author.

produced good tea and served it to outsiders, but the Yunnanese themselves neither understood it nor drank it seriously. He proposed another question: "Shouldn't Puer tea be a calling card for Yunnan? This calling card has been lost by Yunnan and left in alien lands. Now the others come to us, bringing the aged models with them. We'll have to work hard to print out more of our calling cards."

Jiao Jialiang's appeal reflected that Yunnan was, in fact, working hard to "print out" more Puer tea. The 2006 government definition included both raw and artificially fermented tea as Puer tea, despite the never-ending debate over which was more authentic. Once the newly produced raw Puer tea was included in the definition, output increased.[17] *China Newsweek* (Tang Jianguang, Huan Li, and Wang Xun 2007b) cited Zou Jiaju, the vice head of the Yunnan Tea Association, as saying that—according to the 2003 definition, which included only artificially fermented tea—Yunnan's output of Puer tea since the 1970s had been only one to two thousand tons per year; but according to the 2006 definition, which included both raw and artificially fermented tea, the total output of Puer tea reached 80,000 tons per year.

While the fortunes of Puer tea rose, those of other kinds of Yunnan tea declined in terms of both consumption and production. Yunnan used to have several kinds of famous green tea, but I could find none of them at tea trade fairs in Simao, Jinghong, and Kunming in 2007. It seemed that every bit of tea material in Yunnan was being used to make Puer tea (fig. 3.7). I was told by sellers that green tea was less practical than Puer tea because it could not bear long-term storage.

According to a speech in July 2007 by the vice-governor of Yunnan, the output of Puer tea in Yunnan in 2006 reached 80,000 tons, 28,000 tons more than the previous year. Over the period from 2005 to 2006, Puer tea, as a proportion of all Yunnanese tea, increased from 45 percent to 58 percent (Kong Chuizhu 2007).

CONCLUSION: MULTIPLY IMAGINED HOME

Puer tea's home is transregionally authenticated and multiply imagined, embodying what I call its *jianghu,* in which contested desires meet, debate, and negotiate. The packaging of Puer tea has been deeply shaped by ongoing geopolitics.

Scholars in anthropology and political economy have provided similar cases in Europe about the state's participation in promoting profiles of commodities by linking them to locality. This is most clearly exemplified with wine (Guy 2003; Ulin 1996) and cheese (Grasseni 2003) in France, and in relation to "slow food" in countries such as Italy (Leitch 2003). In the early nineteenth century, the French invented the concept of *terroir,* which was supported by, and clarified within, government regulations. *Terroir* stressed that wine or cheese derives certain characteristics from a unique local feature, such as soil and temperature, and also unique local production techniques. This *terroir* cannot be replicated. Furthermore, this concept was also linked to the unique French cultures that shaped the characteristic of the product as well as the inhabitants in the production areas (Barham 2003; Grasseni 2003; Phillips 2006). This emphasis on locality again echoes the notion about authenticity: authentic commodities should contain an aura of originality (Benjamin [1936] 1999). In the case of Puer tea, the state—namely, the government of Yunnan—also participated in linking and emphasizing locality and originality. By updating the definition and production guidelines about Puer tea several times, it tried to identify Puer tea with Yunnan in defense against non-Yunnanese "fake" Puer tea.

In the postmodern era, time and space are compressed as a result of globalization, and the boundaries between spaces become vague (Harvey 1989). However, according to David Harvey, "the less important the spatial barriers, the greater the sensitivity of capital to the variations of place within space, and the greater the incentive for places to be differentiated in ways attractive to capital" (Harvey 1989: 295–296). That is, while all places seemingly share a common global home, each one actively seeks to present a distinct intrinsic self in order to attract investment, commerce, and tourism. In the case of Puer tea, in the process of constructing such a distinction, historical, botanical, and other forms of knowledge were "flexibly accumulated" (Harvey 1989: 295) to authenticate one's unique identification with Puer tea. Compared with the European examples of wine and cheese, the uniqueness of Puer tea lies in the great complexity of forces and the high degree of flexibility in accumulating cultural and economic capital in the ongoing consumption revolution of Reform-era China.

The notion of "translocality" has been used to explain the ongoing mobility and interaction among different places in contemporary China, in which the role of each locality is not weakened and there is "a revitalization of place making and place differentiation," with construction of place identity crossing various geographical levels and "multiple scales" (Oakes and Schein 2006: 2). In the case of Puer tea, the construction of place identity is displayed by the competition between different administrative units about the tea's authentic home. An important strategy of scale is used in these constructions and competitions: while Simao (or Xishuangbanna) declared itself to be the specific origin of Puer tea, it admitted that Yunnan was the general home. Similarly, while Yunnan used Puer tea as a provincial representation, it also stressed that Puer tea embodies the essence of Chinese tea culture. In this strategy, the lower scale becomes the representative of the higher scale. It echoes "ideologies of translocalism" (Oakes and Schein 2006: 7), in which the identification of the local is imaginatively consistent with that of the nation and also of the global, and the economy of the local is practically supported so that it can reach the global standard.

The more different voices try to define a clear boundary for Puer tea, the more complicated and vague its image has become. Such complexity and vagueness, in turn, arouse people's desire to further demystify, reauthenticate, and reimagine Puer tea and its home. In this *jianghu* of Puer tea, these multiple desires have bolstered Puer tea's extraordinary value, but they have also resulted in the impossibility of defining a singular home, a home that can only be multiply imagined.

CHAPTER 4

Heating Up and Cooling Down

Zigong asks, "Shi and Shang, who does better?" The Master says, "Shi has done too much; Shang has done too little." Zigong asks, "Does that mean Shi does better?" The Master says, "Too much is as bad as too little."

—*The Analects*, chapter 10 (Confucius 1981: 184)

When I stayed in Yiwu in 2007, I received many requests from friends in Kunming to bring back some good Puer tea for them. I was happy that more friends were developing an interest in Puer tea, but I also felt uneasy and found this task to be a challenge, since one person's food might be another's poison. My uneasiness grew when I received a call from a close relative who rarely drank tea. He asked me to buy some Puer tea in Yiwu that would increase in value in the future. There were over fifty family brands in Yiwu at that time, and it was hard for me to predict which would increase in value. Several days later he called again, saying that he needed me to buy only a few samples of tea from Yiwu. Another friend in Menghai had collected some famous and expensive teas for him, such as Dayi, the brand of the Menghai Tea Factory (fig. 4.1), and Zhongcha, the brand of the Chinese Tea Company in Yunnan (fig. 4.2).

After I returned to Kunming, my relative showed me the valuable teas that his friend had bought. They were Dayi 7542, which was said to be representative of raw Puer tea; Dayi 7572, said to be representative of artificially fermented Puer tea;[1] and several pieces of Zhongcha, packaged with old paper and declared to be aged. Although these teas had cost him around ¥10,000 altogether, he was keen to obtain more if possible. My relative was busy with his work, and I wondered how he had time to think about tea. He told me that the tea was not for drinking but for investment.

Meanwhile, it became obvious to me that many people were developing a passion for Puer tea. I learned that several of my mother's former col-

FIG. 4.1 Dayi product. The character in the center of the tea cake is *yi* (益), which literally means "benefit" or "increase," and it is enclosed by the character *da* (大), which literally means "big" or "great." Photo by the author.

FIG. 4.2 Zhongcha product. The character in the center is *cha* (茶, tea), surrounded by eight *zhong* (中, China) characters. Photo by the author.

leagues, who used to work in the engineering field, had opened tea shops in Kunming. Near my parents' house, a grocery store on a crowded street was transformed overnight into a Puer tea shop. It seemed that Puer tea was breaking the old Chinese custom of locating tea shops in a quiet place. Even at the vegetable market near my house, I saw an old woman selling caked Puer tea from a portable stall. A local magazine said that the number of Puer tea shops in China had grown threefold from 2005 to 2006 (*Puer Jianghu* 2007a: 15). Even more tea shops opened in 2007.

When the price of Puer tea spiked in the spring of 2007, local people in production areas such as Yiwu stir-roasted tea leaves diligently and often worked until late at night to meet the demand. In Simao and Jinghong I joined in occasional meetings in local tea shops, where regular customers from various occupations gathered, learning tasting techniques and acquiring the newest information on Puer tea. As a Jinghong journalist commented, at this moment, "the entire nation was engaged in tea" (*quan min jie cha*).[2]

People were speculating on Puer tea as they did on stocks in financial markets. In some Kunming teahouses, I saw people drinking and talking about Puer tea around a tea table while checking interest rates and share prices with a laptop. Among this group there was usually a tea expert, who directed members how to properly collect, infuse, and drink Puer tea. There was also often a stock expert, who showed them how to follow the ups and downs of the market. These tea shop groups were made up of people who speculate both on Puer tea (*chao cha*) and on the stock market (*chao gu*).

In Guangzhou, in March and April 2007, the price of Puer tea increased rapidly after it had changed hands in the Fangcun Tea Market. It was said that half the supply of Puer tea in China was in Guangdong, and Fangcun was the biggest national wholesale tea market. Most of the Puer tea there was traded by the *jian*. One *jian* contains twelve stacks (*tong*); one stack contains seven round cakes; and one cake usually weighs 375 grams. So one *jian* has eighty-four cakes of Puer tea, weighing a total of 31.5 kilograms. Before I went to Fangcun, I had heard the saying: "If you ask how much one cake of Puer tea is, no one will pay attention to you, but if you ask how much one *jian* is, the seller will put a chair in front of you and ask for a detailed consultation." In other words, Puer tea was being sold as a bulk commodity and as a product for exchange or storage rather than for immediate consumption. A local trader described the skyrocketing of Puer tea prices in Fangcun at that time:

> The price of some Puer tea varies radically in one day. For example, it is ¥5,000 in the morning, but becomes ¥5,200 in the afternoon. Sometimes it can increase ¥500 in several hours. Not all kinds of tea can be speculated on like this—only Puer tea, as it has the "long-lasting" characteristic.

Before 2007, people had said, "If you don't buy Puer tea when you are young, you will regret it when you become old." During the speculation craze, however, the saying changed to "If you don't buy one *jian* of Puer tea today, you will regret it tomorrow." When my friends learned that I was doing research on Puer tea, many commented that I was making a good choice, hinting that I could gain great advantage by joining in tea speculation (*chao cha*), like many other people in Kunming, Yunnan, Guangdong, and elsewhere in China.

Literally, *chao cha* means stir-roasting tea with fire, often in a very large wok; this is a standard step in rough processing. Stir-roasting tea leaves is seen as one method of killing the green (deactivating oxidation and fermentation in the tea leaves) and boosting the aroma of tea. This technique is said to have become popular in the Ming dynasty (1368–1644).

Different degrees of stir-roasting are applied to different kinds of tea according to different fermentation processes. Green tea is not fermented, and in stir-roasting it one must deactivate all the enzymes to ensure that fermentation will not take place. Oolong tea is partially fermented, and stir-roasting is used when fermentation reaches around 50 percent to stop further fermentation. For Puer tea, stir-roasting pauses the activities of enzymes temporarily but leaves the possibility for further fermentation. Therefore, Puer tea must be roasted at a temperature low enough that it does not kill all the enzymes, but high enough that fermentation is temporarily halted. In Yiwu, I observed that the temperature of stir-roasting is an important issue for both tea peasants and tea traders. The correct temperature is gauged by personal experience rather than actual measurement.

People also spoke of stir-roasting in an extended and metaphorical sense, meaning to deliberately heat and elevate the profile of something, such as speculation on Puer tea. Not all kinds of tea were subject to this sort of speculation, but because Puer tea is considered a "drinkable antique," it could be purchased, exchanged, and stored in the hope that it would increase in value.

The conceptual distinction between "the raw" and "the cooked" may be applied to the transformation of Puer tea. According to that binary contrast, cooking transforms nature and defines culture (Lévi-Strauss 1970, 2008). This is similar to the ideas of maturation and socialization in Chinese and Confucian concepts that refer to processes of acculturation and education. In the classical "culinary triangle," Lévi-Strauss (1970, 2008) compares three cooking methods—roasting, boiling, and smoking—any of which could be more natural or cultural than the others in terms of their cooking conditions and results (see also Leach 1970). In the case of Puer tea, the primary processing method is stir-roasting, which transforms Puer tea from natural leaves into a cultural drink.

So, on the one hand, Puer tea is stir-roasted technically and physically, reaching a balance that is good for drinking as well as for further fermentation. On the other hand, it is stir-roasted metaphorically, heated up toward the humanly endowed values of taste, health, culture, and wealth. Metaphorically, "stir-roasting tea leaves" refers to any symbolic behavior or propaganda, or even to the stock-like speculation that elevated the value of Puer tea.

Both senses of heating demonstrate the binary theory in a particular way: the more artificial interference there is, the great the degree to which the tea is transformed from its original natural features, even though nature has remained an essential component in shaping its superior culture. In the summer of 2007, when the price of Puer tea fell and its values came into question, some attributed its downfall to excessive artificial interference; using the metaphor again, they said that the temperature of stir-roasting had become too high and that Puer tea had become overcooked.

The interplay between multiple human actors, their divergence and interaction in authenticating Puer tea, and their concerns with moderating, criticizing, clarifying, and obscuring the facts about the tea predetermined the fate of Puer tea, from its rise to its fall. And in this regard, the downfall of Puer tea cannot simply be understood from an economic perspective—which mainly attributed the recession to speculation, greediness, and failure to obey basic economic rules—but has to be understood in terms of the public debate about cultural values.

TWO EARTHQUAKES

In June 2007, an earthquake of 6.4 on the Richter scale occurred in Simao (subdistrict). Its epicenter was in the town of Puer, which had been renamed Ninger just two months earlier. Three people were killed, more than five hundred were injured, and over a million suffered property damage in Puer. The direct economic loss was ¥25 billion (CCTV 2007a), and people were very concerned about the impact on Puer tea production. One economic analyst, Xia Tao, expressed this concern in a television discussion: "The price of Puer tea had increased throughout the spring. Theoretically it should have jumped higher due to the production loss caused by the earthquake. However, the price didn't rise after the earthquake. That's when I said to myself that something was going wrong" (CCTV 2008).

An economic report on Puer tea by China Central Television Station 2 was screened on June 15 (CCTV 2007c). This report was taken as the turning point after which Puer tea's profile suddenly diminished—though the reduction in price had actually begun earlier—and it became a specific target that Puer tea supporters later challenged. The report stated that two earthquakes were affecting Puer tea: the earthquake in Puer City (renamed from Simao), the production area of Puer tea; and the earthquake in the marketplace. Titled "The Bubble of Puer Tea Is Broken," the thirty-minute program opened in the Fangcun Tea Market in Guangzhou, whose Puer tea price was acknowledged to be the barometer for all the tea markets in China. The report said that the price of Puer tea at Fangcun had fallen by half in the previous thirty days. A *jian* of 7572 Puer tea (31.5 kilograms) that had sold for ¥20,000 the month before was now worth only ¥9,000, and according to the report, its factory price was only about ¥5,000. Drawing a comparison with speculation on the stock market, the report described the large tea factories and companies, along with their distributors, as an invisible hand "shuffling the tiles." It gave an example to show how, in the previous few months, Puer tea's value had been deliberately elevated:

> Let's take 30 kilograms of Puer tea as an example. Its factory price is ¥4,800; the first-level distributor, who acquires the dealership at a very high cost, sells only 20 percent of his stock, which causes misguided information that this product is scarce in the market. The dealer then repurchases the tea at a higher price to elevate its value and sells all of his stock at this higher price. After this speculation is repeated by the second- and third-level distributors, the price of this Puer tea reaches ¥23,000. The private investors, who buy it at this stage, have been deeply caught up in the market, and it is hard for them to get away.

According to the report, the rising price of basic tea material in the production area was also spurred by the deliberate elevation in urban markets. And when the most powerful "bankers"[3] suddenly withdrew, the private investors and middlemen could do nothing but cry.

The report included several amazing figures:

1. The fee that the distributors must pay to the big tea factories and companies for distribution rights ranged from ¥1 million to ¥30 million. The more they paid, the more Puer tea that they could order. According to the report, this was the key reason that Puer tea's price had become so extraordinarily high, since all the distributors had to find ways to recover their initial investment.
2. Dealers had between 100 and 300 tons of Puer tea in stock that they could not resell since the price had collapsed.
3. Almost all tea products (95 percent) were bought for speculation and storage, with only 5 percent purchased for actual consumption. One tea expert in Guangdong said that "even without buying another piece of Puer tea from the production area, Guangdong could not drink all of its stored tea for five to eight years."

The report drew the conclusion that Puer tea's rise and fall was due to improper speculation. It asked ordinary investors a rhetorical question: "If the quantity of a certain commodity in storage is far more than its actual consumption, would it be valued so high?" The report had great repercussions due to its national broadcasting impact. It became the hottest topic for discussion among people who cared about the Puer tea business.

Another influential media report appeared in the magazine *New Generations* (Xinshengdai), published by the Sichuan Youth Newspaper, which said that the number of people in China involved in Puer tea speculation had reached thirty million. It argued that this large population had been tricked and that bankers, illegal traders, large tea companies, and government officials were all complicit (Guo Yukuan 2007a).

Like the CCTV 2 report, the *New Generations* article also examined Puer tea's popularity as a focus of stock market–like speculation, taking as a case study the problematic business model of Zhongcha, the Chinese Tea Company brand in Yunnan. The Zhongcha Company had authorized private companies to "produce" Puer tea that was specially packaged as Zhongcha. What these private processors actually sold was their label, not any assurance of authentic technique or quality. As a result, fake Puer tea flooded the market (Guo Yukuan 2007a). Like my relative who asked me to collect Puer tea for him, many investors spent their money on brands like Zhongcha that were reputedly produced by famous Puer tea companies, thinking that these brands would increase in value. The actual downturn in the tea market, however, made it hard for them to resell their teas, and the report by *New Generations* further swayed people's faith in the authenticity of these products. The tea that my relative bought was, as far as I knew, still untouched. But these were just minor cases. There were additional reports about a more serious crisis among larger distributors.

Media reports on Puer tea had flourished ever since the tea trade had begun to pick up. By 2007 there were thirteen Puer tea magazines, ten of them from Yunnan. Most had been established between 2005 and 2007, in print, television, or web form. There were also special columns in newspapers devoted to Puer tea (*Yunnan Puer Cha* 2007). Before the bubble burst, these media reports, especially those from Yunnan, had always carried "good news" about Puer tea, celebrating its value and reporting on events like Simao's name change and the new tea caravans. By contrast, national media, like CCTV 2, and media from other provinces, like *New Generations,* became the pioneers that spread "bad news" about Puer tea. In the face of recession, for the first time some of Puer tea's assets were deemed neutral or perhaps even negative factors. The production process, which had once been described as delicate and labor intensive, was now reported to be "not complex and mysterious at all" (CCTV 2007b), and the tea's medicinal function, which had been positively publicized before, was now pointed out to be "almost the same as other sorts of tea" (CCTV 2007b). One report said that "the special function of Puer tea still awaits further proof" and that "propaganda that Puer tea has a long-lasting quality is misguided" (Wang Xun 2007; CCTV 2007d).

THE VOICES FROM YUNNAN

I attended a symposium in Kunming at the end of June 2007, two weeks after the show about Puer tea's recession had been broadcast. The symposium was initiated by the Puer Tea Association of Yunnan and was attended by state government officials, local media, principals of several big tea companies, tea experts, and some self-invited tea traders. Most of the attendees were from Yunnan, though a few guests were invited from Beijing. A banner displayed the title of the conference: "Voices from Yunnan—The Clarification of Puer Tea's Current Situation." The atmosphere was very serious, and the symposium sponsor adopted an attitude of hostility toward the central media report. It said informally that the CCTV 2 report was the result of bribery by groups who were hostile to Yunnan's Puer tea and that the symposium was a counterattack. *Puer Tea Weekly,* a weekly newspaper supported by the Puer Tea Association of Yunnan, described the purpose of the symposium:

> Puer tea has encountered fluctuation recently and its sales are at a temporary standstill. Some false reports, together with spiteful rumors, have seriously harmed this newly emergent industry. These are the conditions under which this symposium was organized. (*Puer Tea Weekly* 2007a)

The "false reports" mentioned in this statement referred to the claims by media outlets such as CCTV 2 and *New Generations.* Some attendants at the symposium expressed their anger that the key feature of Puer tea—its longevity—was being undermined by these reports, and that the tea's medicinal function was being compared to that of other kinds of tea. One speaker surmised that these "negative" reports might have been plotted by groups whose interests had been harmed by Puer tea's popularity. An article published in the *Kunming Evening News* echoed this idea, and the author recalled several events in which Puer tea had been deliberately attacked, including a 2004 newspaper article that had said that not all Puer tea was worth storing, a 2005 story "created" by the Guangdong media that some Puer tea was fermented in a pigsty, and a 2006 report that twelve kinds of Puer tea produced in Yunnan were not quality products. The author of this article concluded that Puer tea had frequently been attacked by false or exaggerated reports. The falling price in 2007 actually affected only certain brands, such as Dayi and Zhongcha, but in media reports it had been described in exaggerated terms as the collapse of the entire Puer tea market (Lu Ming 2007).

At the end of the symposium, despite some dissenting opinions, a manifesto was composed, calling for the "protection of Puer tea" and "defending Puer tea against any harm." The manifesto also stressed that Yunnan was the original home of Puer tea, and that Puer tea had to be made from Yunnan's large-leaf tea variety.

Soon after the symposium, in July 2007, the Tea Association of Yunnan launched the Puer tea geographical trademark. A production or trade unit could put the trademark on its tea products only after it had been certified and had paid certain fees.[4] The function of the trademark was to assure the authenticity of Yunnan's Puer tea (*Puer Tea Weekly* 2007b).[5] But it was not a compulsory regulation, and only a small number of traders registered to use it.

In the following months, the Tea Association of Yunnan and several allied tea units organized a series of activities. Mass media outlets from Beijing were invited to Yunnan to report on the "truth" about Puer tea. At the same time, a panel from Yunnan went to local tea festivals and trade fairs in Beijing, Tianjin, and Shanghai to publicize Puer tea's positive image (*Puer Tea Weekly* 2007c). In November, the Yunnan Provincial Government organized more panels and encouraged private tea companies and traders to participate in the Tea Trade Fair in Guangzhou, the most important annual tea fair in China. Several of my informants who had attended this trade fair reported that other kinds of tea, such as oolong tea from Fujian and dark tea from Guangxi, were being promoted at the trade fair as being suitable for long-term storage. As one person said, "These people of other provinces criticized us, saying that we can't say that Puer tea could be stored for ages, but in fact they themselves were utilizing this long-lasting feature."

When the long-term value of Puer tea came into doubt during the downturn, people in Yunnan began to advocate that it was time to drink rather than store the tea—that unless people did so, its value could not be fulfilled. This was clearly an attempt to address the claim on CCTV 2 that it would take more than five to eight years for Guangdong residents to drink all the Puer tea that was stored.

OVERCOOKED OR TOO RAW?

Although the provincial government had developed a series of events to "protect" Puer tea, criticism continued, both formally and informally, attempting to reflect upon the root cause of the recession. Multiple public opinions circulated, illustrating the cultural characteristics that shaped the rise and fall of Puer tea.

Some said that Puer tea's downfall was due to excessive propaganda; metaphorically, it was overcooked. Others attributed the crisis to a lack of proper effort; metaphorically, it hadn't been cooked enough and was still too raw.

For those making the "overcooked" argument, the government, whether at the provincial or district level, was the main culprit. Since 1993 there had been over ten conferences on Puer tea (some attached to trade fairs) inside and outside of Yunnan, all sponsored by local authorities. In April 2007 there were three in rapid succession, separately organized by the district governments of Simao, Xishuangbanna, and Lincang. The "welcoming" of the Golden Melon, Simao's name change, several caravan expeditions, the linking of Puer tea to the *Götheborg,* various celebrity performances, the auction of aged Puer tea, and even the elevation of the tea price were all criticized as deliberate overpackaging of Puer tea by the government. For instance, while the 2005 caravan from Puer successfully finished its trip, and the Puer tea it carried was auctioned at an amazing price in Beijing, the other caravan, setting out from Yiwu in 2006, was considered a failure and an ill-planned repetition. Those in Yiwu who had contributed Puer tea to this caravan were unhappy, because nobody told them where their products had ended up. In fact, due to a shortage of funds this caravan had to pause in Zhejiang (near Shanghai) before heading toward Beijing. While the participants were looking for further funds to continue the trip, their horses were stolen and they had to sell the remaining tea to cover the cost of their transport home. As Huang Bingsheng, the vice head of the Standing Committee of Yunnan Provincial People's Congress, commented, "Good things don't happen more than three times" (*haoshi bu guo san*). His words reflected a view that success could not be guaranteed, especially when it was based on imitation (CCTV 2008).

The mass media was also criticized. According to one representative at the "Voices from Yunnan" symposium, exaggerated propaganda by various mass media outlets had been detrimental for Puer tea. He thought it was improper to publicize forest tea's superiority over terrace tea, for instance, or to exaggerate Puer tea's medicinal function. He believed that the price of Puer tea would have gone down if the tea had not been promoted by the media as a stock for speculation.

When participants in the symposium began to discuss the "stock market value" of Puer tea, I suddenly realized that the meeting room we were in, situated in a wholesale tea market in Kunming, was the actual place where electronic transactions for Puer tea took place. The screen, which that day showed the CCTV 2 report, was normally used to display the latest information about Puer tea, such as the most recent price of a particular tea product. It functioned just like the electronic screen that displays interest rates in a bank.

Critical commentators said that there were other kinds of "overdone" behavior; they blamed large tea companies that asked for excessive distribution fees and approved inauthentic production, skillful bankers who purposely tricked individual investors, and big traders who were hostile to Yunnan's Puer tea. Individual investors and consumers, who had been too greedy and now encountered bankruptcy, received both blame and some sympathy.

Professor Di, an expert in microbiology, regarded it as inappropriate that the tea scholars of Yunnan had attempted to extricate Puer tea from the six established categories of Chinese tea. When I interviewed him he said that he thought it would still be better to include Puer tea in the category of dark tea, since they shared many similarities in production procedures, although not all. By way of comparison, he said that people were human beings first, and then could be further classified into certain races. Removing Puer tea from the established classification system, Di said, had isolated it and made it vulnerable to attack. Speaking from experience, he told me that dark tea produced in Hunan and stored for a long time was also drinkable. Therefore, he said that longevity was not a unique feature of Puer tea, but just that it had been overemphasized.

Some comments I heard in the teahouses echoed Di's point of view. Tea drinkers complained that Puer tea's supporters should not have criticized other teas when promoting Puer tea. This, they said, had violated a rule of advertising and damaged the profitability of other teas. Criticizing the aggressive promotion of Puer tea, many people cited old Chinese sayings, such as: "Tall trees catch much wind" (*shu da zhao feng*) and "The bird that stands out is easily shot" (*qiang da chu tou niao*). To them, Puer tea's demise confirmed these commonly held old beliefs. In traditional Chinese philosophy, originating from Daoism and Confucianism and applied in daily life, the top position is dangerous because it is envied and easily attacked by others. This recalls the doctrines that are often suggested to a knight-errant wandering in the risky *jianghu:* he should not show off too much; he should hide his real thoughts and look humble even if he is a skilled martial artist, lest he attract too much attention and suffer from his distinction. When people reflected that Puer tea had been overspeculated, overpropagandized, and inappropriately redefined, they were suggesting that Puer tea had failed to obey a basic Chinese concept and was therefore attacked.

Some also held the opposite opinion: that Puer tea hadn't been cooked enough and was still too raw. At the "Voices from Yunnan" symposium, a special guest from Beijing appeared, wearing sunglasses. He had never shown his face to the public, but his name, Wang Hai, was famous. He was known as "the pioneer of cracking down on counterfeits in China" (*zhongguo da jia di yi ren*). According to him, the report by the central media was true, and the crisis lay in the unclear value of Puer tea. He raised a series of questions: Was Puer tea a beverage or a medicinal tonic? Was it really a "drinkable antique"? What was fake and what was authentic? How could the exact age of a cake of Puer tea be known? Why didn't people report it when they chanced upon the production of fake tea? Was there a clear and scientific regulation to supervise all of these issues? He hinted that the answer to all the questions was "uncertain." As everyone knew, these problems had been neither solved nor addressed by the authorities, the tea researchers, the tea companies and traders, or the consumers. As a result, identification of Puer tea usually depended upon individual experience.

Wang Hai's speech and other similar statements called for clearer and stronger regulations on Puer tea. Likewise, the propaganda sponsored by the state and implemented by the mass media was thought to be insufficient; the publicity promoting Puer tea, according to some commentators, hadn't been related well to the "authentic" culture of Yunnan and hadn't sufficiently shown the contributions of multiple ethnic minorities; and it was not that Yunnan had produced excessive tea but that the market didn't have enough authentic Puer tea.

These points of view, regarding "inadequacy" and "rawness," actually responded to the "overcooked" argument in another way. They reflected the fact that the output of Puer tea and cultural packaging around it was increasing in quantity, but that not many of these products were authentic. They highlighted the importance of quality rather than quantity. When quality was bad, quantity was a waste. They implied that Puer tea had been overcooked as well as too raw, and they asked for a proper method to accurately identify the quality of Puer tea.

In fact, appeals for accuracy and more powerful regulation had been made even before the Puer tea recession, but they had never been successful. For example, at the laboratory of a professor in an agricultural university, I was told that a scientific method was being developed to shape a specific flavor for Puer tea. This professor proposed "digitizing Puer tea" in order to artificially and quantitatively control the process of Puer tea production. However, there was great opposition to the development of such a scientific approach. At many teahouses, I heard traders and consumers saying that the charm of Puer tea lay in its endless variation; you never knew what a particular piece of Puer tea would taste like, and any attempt to fix or "digitize" it would be useless. A tea trader from Hong Kong responded sharply, telling me that he believed a quantitative method for establishing Puer tea's authenticity could be developed, "but then, you can imagine how boring the process of appreciating and distinguishing tea would become!" This man is considered a "super" tea expert. He tastes many tea samples each day to make decisions for his business. He told me that he tasted so much tea that, one time, he experienced stomach pain and had to go to the hospital for an injection. After recovering, he paid more careful attention to how much he drank but still continued to enjoy his own sensory tasting.

In Xishuangbanna, a local governmental official told me that he wished the origin of tea material could be ascertained from the textural characteristic of the tea leaf, like the method used in police investigations to trace somebody's footprint or handprint. In this way, he said, people could identify whether the tea material was from Yiwu or Menghai, and whether it was made of forest tea or terrace tea. This idea was opposed, too. The opponents were not concerned about whether such a "scientific" method could actually work in technical terms, but they worried about whether it could be successfully implemented, as exemplified by the comments of another local official. This official expressed his concern very directly to me: "Even if there were a way to tell where tea material had originated, I think after all it would be up to the key person involved to divulge where it was from."

Once a proposal attempting to more clearly define and distinguish Puer tea was put forward, oppositional voices emerged, arguing for a return to the original vague "raw" situation. Their concerns largely obstructed the implementation of all the proposed revolutionary ideas. Facing such endless debates, I wondered whether vagueness itself was a prerequisite for Puer tea in the cultural and social context of China.

CONCLUSION: A CULTURAL DILEMMA

Like Puer tea's rise and fall in Yiwu, the whole Puer tea industry in Yunnan suffered from a cooldown after heating up. Blame was extended in several directions, and past efforts that had successfully promoted the high value of Puer tea were considered to have led to indigestion in the entire market. Economic analysis coming out soon after the recession, as exemplified by the CCTV 2 report, attributed the market downfall to excessive speculation, improper investment, blind zeal, and greed. However, this economic analysis was regarded as false and even as one of the factors that accelerated Puer tea's downfall. Following this, additional voices emerged to reflect upon the situation.

In the debate about Puer tea's downfall, there is a continuing concern with "moderation," one of the doctrines in Daoism and Confucianism that is embodied in everyday life. Food should not be cooked too much, nor should it be too raw, and everything must find its proper place, a middle ground. This basic belief is applied to people's behavior, their relationships with others, and their understanding of objects. In diagnosing the cause of the Puer tea recession, there was a strong element of self-criticism in public opinion. That is, Yunnan's Puer tea came under attack because Yunnan itself, represented by the provincial government, had "overcooked" Puer tea and damaged the interests of others. Under the doctrine of moderation, there was also a concern with equality and compromise, exemplified by the updated definition that identified all production areas in Yunnan as the home of Puer tea.

This belief about moderation was challenged by the new culture in Reform-era China of self-presentation and profit accumulation. Even after the recession, and faced with the "overcooking" criticism, the provincial government still attempted to celebrate and promote Puer tea. Any criticism of Puer tea was taken as negative, hostile, and contrary to the interests of Yunnan. Truth or nontruth was irrelevant. Only profit mattered.

There were also revolutionary voices appealing for an immoderate and precise image for Puer tea. These voices asked for stronger, clearer, and more scientific regulations and supervision, saving the tea from being "too raw." They hoped that through these methods Puer tea could become clearly identified. These appeals, however, encountered difficulties even before they could be put into practice. As a result, the "rawness" and vagueness continued, and attempts to define Puer tea diverged.

These debates showed that the desire to package Puer tea and the alternative desire to unpack it had coexisted long before the actual recession. The contest between them had largely shaped the story of Puer tea from its heating up to its cooling off, and speculation became the fuse that hastened the transformation. The debate goes on endlessly, and so both in its rise and fall, Puer tea's authenticity remained complex, multifaceted, and vague. Even moderation, one of the strong themes in public opinion, is not measurable by any quantitative data, but it is more dependent on personal experience and interpersonal negotiation. The boundary between what is overcooked, what is still raw, and what is nearing moderation, is also contextually determined. Tradition encourages people to remain in the middle ground, but they are taught in the new era to present themselves effectively; they appeal for clarity, but in fact they enjoy vagueness. Thus it becomes a cultural dilemma for actors to decide to what extent Puer tea should be cooked.

秋 AUTUMN 愁

Yellow Leaf Puer tea (Lao Huang Pian) in Yiwu.
Photo by the author.

CHAPTER 5

Puer Tea with Remorse

One always fears the coming of the Moon Festival,
For flowers and leaves are withering.
Rivers flow east to the sea.
When can they flow back again?
If one does not work hard in his youth,
He might mourn vainly in his old age.

—*Long Song* (Chang ge xing), anonymous,
Han dynasty (202 B.C.E.–220 C.E.)
Translation by Ch'en Chao-ying,
in Ho Chi-p'eng (1995: 354)

I returned to Yiwu in September 2007. It was mid-autumn, the other important season for tea harvest. Taking the bus from Jinghong, I arrived in Yiwu in the late afternoon, just as I had in the spring. But something had changed. The main street was obviously quieter. The grocery stores and restaurants were open as usual, but with few customers the owners looked idle. Learning a lesson from my experience in the spring, I had booked a room with the guesthouse. But when I arrived, I found that there were no other guests.

Since the beginning of autumn, some locals who had business alliances with outside traders had been to Jinghong, Kunming, or Guangdong to find out what was happening in the urban market. They had brought back the important information that a recession was evident in the Puer tea market everywhere. During the spring, the price of *maocha* in Yiwu had been high, around ¥400 per kilogram for forest tea, and many traders were coming to compete in the trade. But by autumn, the price had fallen to around ¥100 for the same kind of tea, and few traders came to buy.

Having enjoyed the rising price in spring, people in Yiwu were frustrated by the big contrast in autumn. Suspicions arose about the change in the tea price and its future development. Mr. Guan, a seventy-year-old man, asked me to tell him more about the situation of Puer tea in different urban areas, as he could not visit them personally. He asked me a serious question: "Do you think the price of our tea will go back to the low level it was before?" He was referring to the period from the 1950s to the 1980s, when tea material in Yiwu was worth less than ¥5 per kilogram. The question initially sounded absurd because there was such a big gap between the ¥5 of old and the more than ¥100 or even ¥400 that just one kilogram of tea had recently commanded. When the price was only ¥5, tea was insubstantial and even despised as "negative capitalism," especially during the Cultural Revolution (1966–1976) (see also Zeng Zhixian 2001: 93). If the tea price went back to ¥5 per kilogram, it would imply that Puer tea was again valueless and negative. How could this happen?

When Mr. Guan asked me this question, we were sitting in the courtyard of his house. The house had been built recently and had cost almost ¥250,000. It was a two-story modern brick and concrete house with an antique-style finish on the railings. From the 1950s to the 1970s, his family had lived in a thatched shed; in 1987 he built a tile-roofed house; in 1992 he built a simple brick house for ¥50,000, which was later used as a guesthouse; and in 2005 he built a new brick house for ¥180,000, which he sold later in order to build the present one in a new location. In the center of the courtyard was a car, which was said to have cost ¥130,000. That Yiwu families were able to afford to build such houses and to buy such cars was mostly attributable to the soaring tea business of recent years. The development of Puer tea had catalyzed change in almost every corner of Yiwu. During my fieldwork I often heard tea drinkers say that tea couldn't be equated with rice, implying that eating basic foodstuffs is more important than drinking tea, no matter how valuable the tea. But for people in Yiwu, tea had become equal in value to rice, and they were eating fully and living better because of it. Observing the anxious expression in Mr. Guan's eyes, I began to think about the reasons for his concerns.

The ancient Chinese philosopher Mencius famously said, "Appetite for food and sex is nature" (*Shi se xing ye*). After the Reform era, when Chinese people became wealthier, this statement was taken for granted as true. But during the Maoist period (1949 to the late 1970s), when China experienced hunger and when class struggle was deemed more important than eating, "the existence and indulgence of non-collective appetites were almost an embarrassment" (Farquhar 2002: 3).[1]

The contrasting attitudes toward eating in the Maoist period and since the early 1980s show that indulgent consumption in contemporary China may be attributed to the unforgettable shortage of food in the past. The present repletion, whether for the individual or for the nation, can be diagnosed in terms of past depletion.[2] In this regard, Mencius's saying is not always true but "timely" (Farquhar 2002: 2).

Linking past experiences to present orientations is useful for understanding the situation of Puer tea. Local attitudes toward Puer tea have long been shaped by domestic policy and wider external impacts. Just as Mencius's saying should be looked at in context, the present value of Puer tea in Yiwu should not be taken as natural; rather, it should be reread flexibly as "responding to the specific character of place, time and person" (Farquhar 2002: 108). Nor are local production and consumption of Puer tea shaped by a fixed custom. In fact, they have changed back and forth as a result of many unexpected factors. As a result, the root causes of the present worries are not only found in the recent rise and fall of the market. They also need to be looked at through the shadow of the past. And "worries" not only refers to worrying but also contains more reflective thinking about the vicissitudes of Puer tea. As Sherry B. Ortner (2006: 11) argues, "History is not just about the past, nor is it always about change. It may be about duration, about patterns persisting over long periods of time."

A BRIEF HISTORY FROM NATIONALIZED TO PRIVATE TEA BUSINESS IN YIWU

In popular books on Puer tea and in local stories about Yiwu, the historical period from the mid-eighteenth to the early twentieth century is described as a time when private tea companies prospered (Zhang Yingpei 2006; Zhang Yi 2006a; Ruan Dianrong 2005a; Deng Shihai 2004). People enjoyed recalling stories about how the tea business was extended to other Southeast Asian areas and how tea became the dominant element in local livelihoods. At that time, little rice was produced in Yiwu; instead, it was obtained via exchange for tea (Jiang Quan [1980] 2006: 46).

After the late 1930s, Yiwu's private tea business suffered from war and other turmoil, and even after the situation stabilized, it still struggled to revive. Soon after the foundation of the People's Republic of China in 1949, privately owned tea companies became nationalized in the early 1950s. Tea was brought under a state monopoly on buying and selling. After rough processing, *maocha* produced in Yiwu was carried to the national tea factories outside for fine processing.[3] This arrangement lasted almost half a century, until the private tea business was revived in the late 1990s.

Contemporary popular books on Puer tea rarely mention the period when the tea business was nationalized, and few people in Yiwu actively talk about it, despite their talent for telling stories about the prosperous period that preceded it. For many locals, this was simply the period when Yiwu became a mere supplier of basic tea leaves for national tea factories, and telling such an inglorious story would do nothing positive for today's tea development. Mr. Guan's anxious question and the low prices paid in the past, however, made me realize that this period was never actually forgotten. Together with the previous period of historical glory, it was rooted in local people's memory as an important reference point when they considered their current and future livelihood. During this period four notable aspects of Puer tea changed in Yiwu: the relationship between tea and rice; the planting methods; the definition of Puer tea; and the price. All of them had changed according to the government's variable tea policy.

First, from the 1950s to the 1980s, the development of tea production took place within the context of the high priority placed on food production. Yiwu is located in a mountainous area with an average altitude of 1,300 meters. In the past many highland areas had long been used for growing tea, while the flatter lowlands were used for planting rice, corn, or legumes. However, during the food shortage from the 1950s to the 1980s, many highland areas were converted to food production. "Planting food as a guiding principle" (*yi liang wei gang*) was the slogan for most of this period, when the area was deficient in basic consumer goods. In local people's memory, the time from the 1950s to the 1970s was the "hungry period," when at least one-third of local families were grain-deficient households (*que liang hu*) that had to survive by relying on minor cereals and relief food. The early 1960s was the most difficult time, as China was suffering from a disastrous famine and an economic crisis. As Mr. Guan told me, in those years half of the local food grain had to be handed to the state, and with little remaining food many people had to survive by going to the mountains and collecting taro or wild vegetables. Locals attributed the slow development of rice production to the inefficient collective working system and the lack of technology at the time.

Tea production developed slowly, too. When food was in very short supply, tea trees in some areas were cut down and the fields were converted to rice production (see also YTIEC 1993: 20). Tea farmers worked less actively due to the low value of the tea, and at times the money they earned from tea was so bad that it could "only provide income to buy salt and pepper," as one Yiwu resident told me. Despite this situation, tea was still the main source of income in Yiwu. The paradox was that the development of tea had to defer to food production, but food was still in short supply, even with some income from tea.

Second, although food production had been kept as the focus, during several periods tea production was emphasized and boosted for specific political reasons. In 1958 the Great Leap Forward stressed quantity rather than quality of production. Encouraged by blind enthusiasm, people worked too hard, and many tea trees were overpicked (see also YTIEC 1993: 20).

The next upsurge was in 1974, responding to the request to improve tea areas by the national conference on tea held that year (YTIEC 1993: 22). But because this was during the Cultural Revolution (1966–1976), political movements were prioritized over tea production, and tea production did not develop much (see also Etherington and Forster 1993).

After the Cultural Revolution ended and when China began the policy of Reform and Opening Up (the late 1970s and early 1980s), a new system of cultivation had a greater impact on Yiwu's tea production. The food deficiency problem was relieved, and the government began to encourage people to work harder on tea. In the 1970s, tea fields were reallocated from the collectives to private ownership. In the early 1980s, tea seedlings were brought in from other tea-producing regions of Yunnan, such as Lincang and Jiangcheng. Local peasants were encouraged to plant them without pay (see also EBMCA 1994: 227). In 1982, the inhabitants of a subvillage in Yiwu became full-time tea peasants (Mengla Archive 1982). As for the method of planting, a new mode of arranging tea in regular and dense terraces was advocated in order to boost output.[4] This new method of cultivation was considered scientific and advanced because it made it easier for farmers to manage the tea fields and increase tea output. Zhang Yi, who worked in the Yiwu government at that time (he later became the pioneer of the

FIG. 5.1 *(above)* Terrace tea is easier to pick. Photo by the author.

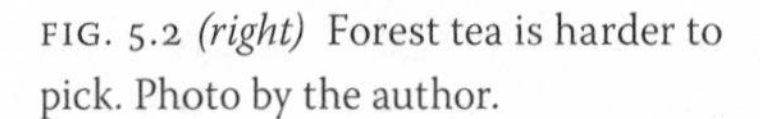

FIG. 5.2 *(right)* Forest tea is harder to pick. Photo by the author.

revived private tea business in Yiwu in the mid-1990s) took a team to study the scientific way of planting from neighboring tea regions in Menghai and then popularized it in Yiwu. The new terrace teas were trimmed regularly to maintain their bush form and to prevent them from growing as tall as the older tea trees, which were the actual botanical form of tea (figs. 5.1–2). Tender tea buds rather than rough tea leaves or tea stems were appreciated. Meanwhile, old tea trees that were over two meters tall were pollarded (figs. 5.3–4).[5] They had been grown for generations, scattered in the forest, and become tall and low yielding. Cutting them short encouraged rapid regrowth and made picking more convenient. At that time, these new forms and operations were taken as scientific methods, whereas the old ways of forest planting were regarded as primitive and backward. The differences between terrace tea and forest tea, and between pollarded and nonpollarded forest tea, did not become important until the resurgence of Puer tea in Yiwu at the beginning of the twenty-first century.

The third important feature of this period of tea nationalization was that

FIG. 5.3 *(above)* After being pollarded, tea trees produce many new branches and multiple trunks. Photo by the author.

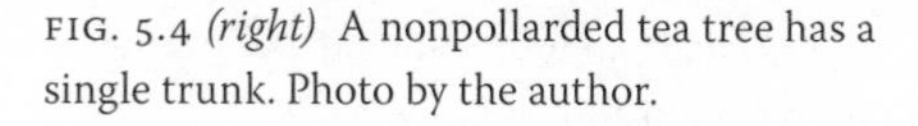

FIG. 5.4 *(right)* A nonpollarded tea tree has a single trunk. Photo by the author.

the definition of "Puer tea" was vague, unimportant, or even lost for Yiwu people. Talking with many locals, I found out that they seldom regarded the tea produced during the nationalized period as Puer tea; instead they called it "sun-dried basic tea leaves" (*shai qing maocha*). This *maocha* was sent to national tea factories for fine processing, a process that Yiwu people no longer had to undertake themselves. This *maocha,* in people's earlier conception, was not Puer tea because it was loose rather than pressed, and people drank it as fresh as possible, rather than after years of aging. Once the *maocha* turned bad, they threw it away, unlike today, when people store tea as a proud investment. I was told by locals that the practice of long-term storage didn't come to Yiwu until after it was "rediscovered" by the Taiwanese in the mid-1990s. What was ironic and confusing was that people prepared *maocha* for the national factories the same way that they prepare *maocha* for their own private businesses today. But they regarded the former as being more like green tea, while the latter was declared to be Puer tea, or at least it was called the basic tea leaves of Puer tea.

Fourth, the tea price increased very slowly from the 1950s to the early 1990s. As figure 5.5 shows, it took almost ten years for the tea price to increase from ¥0.20 (1950) to ¥1.00 (1959) per kilogram, and an additional thirty years to reach ¥10.00 (1992). Xu Kun, the head of the Xishuangbanna

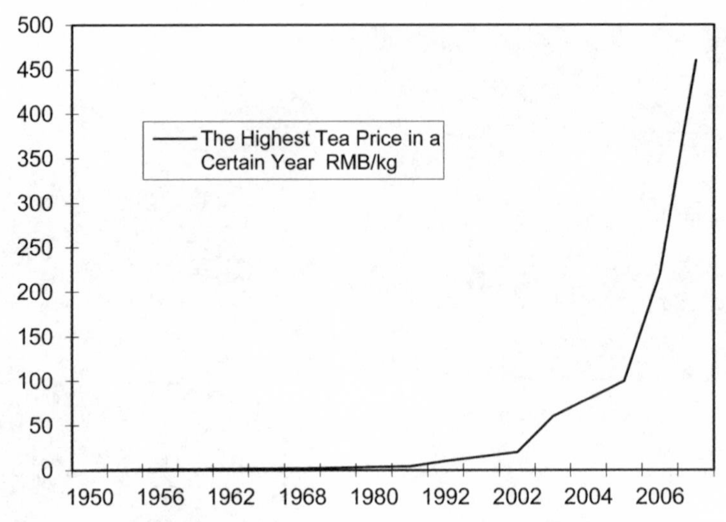

FIG. 5.5 The price of tea in Yiwu, 1950 to 2007. Data from 1950 and 1990 is derived from YTIEC (1993: 69–71); data from 1986 to 2007 is based on fieldwork interviews.

Supervision Bureau of Technology and Quality, recalled that in 1992, when the tea price in Xishuangbanna was only ¥10 per kilogram, he was visited by a special guest from Hong Kong. This guest suggested that the local tea production should be placed under better supervision because he had witnessed how "lazy" the local peasants were when they worked in the tea fields, and he felt it was a pity that the good tea material was not being well managed. Since the early 1990s, people from Puer tea consumption areas, such as Hong Kong (and, later, Taiwan), had been attracted by the value of Puer tea and started paying attention to the situation in the production area. Though the Hong Kong guest's impression about laziness might well be a prejudiced view of local production techniques, local people hadn't started actively working on tea because tea prices were still low. At present, however, outside traders were worried that some locals were working too hard, overpicking the trees and curbing the tea's natural growth.

FOOD OR TEA

Every morning between seven and eight o'clock, there was a busy open-air market on central Yiwu's main street, with vendors selling all kinds of foodstuffs. At the market I often met Mr. Zheng, who came to shop there daily. Like most locals, his family didn't own a refrigerator, and he had to buy basic foodstuffs for cooking every day. Before starting his Puer tea business in 2005, Mr. Zheng didn't come to the morning market every day, or at least he didn't need to buy as much as he does now. Like most people in Yiwu, he used to raise pigs and chickens, cultivate rice fields, and plant vegetables for household use. But in recent years the Puer tea business had taken more and more of the family's time and energy, and all other activities had to cede their position to Puer tea. The Quality Safety Standard had also set strict rules for tea processing, leaving less space for livestock. Although some people running tea businesses tried to keep several pigs, they butchered them only on special occasions and went to the market for daily needs. At Mr. Zheng's house, there was now not a single hen or pig. Even the family production of soy sauce, a custom that Yiwu people had inherited from their ancestors in Shiping in southeast Yunnan, had stopped, as Mr. Zheng worried that the strong smell of soy sauce would affect his tea. Therefore, for Mr. Zheng's family, most provisions for eating had to be obtained from the market, leaving Puer tea as the only notable product made at home.

Tea, more than anything else, was helping local families live better lives. Over four months in the spring of 2007, I witnessed or participated in numerous rural family banquets (*sha zhu fan*) that involve slaughtering a pig and inviting relatives or close friends to share in the feast. The pig, for many people engaged in the Puer tea industry, was usually bought from the market rather than raised by the family. In the past, when rice was insufficient, banquets like this were rare even during festivals; ten years ago, such banquets would have been possible in Yiwu only during the Chinese New Year or at a wedding ceremony. But now, many families held more than one banquet each year. The obvious reason for this was the rising income from tea (fig. 5.6).

After becoming engaged in the Puer tea business in recent years, many families had abandoned their rice fields. In October 2007, I conducted a survey of twenty-three families living on the old street of central Yiwu. All were involved in tea production, but only two families were still growing rice. The desolate rice fields were unused, converted to tea cultivation, or rented to immigrants from other ethnic groups. Tea had supplanted rice as the mainstay of local livelihoods. As in the time of their successful ancestors, present-day Yiwu people ate rice imported from other regions, such as Menghai, purchased with their income from tea. They drank tea when

FIG. 5.6 Dishes at the wedding banquet. The one on the bottom right is tea. Photo by the author.

their stomachs were full. Like many urban tea drinkers, or even like the royal families more than a century ago, they drank tea to counteract greasiness.

These lifestyle changes would have been accepted by Yiwu people as a normal phenomenon if the Puer tea trade had continued to expand. However, the recession in the Puer tea market cast a negative shadow over their prosperity. Many local people compared the downturn of Puer tea with the continuously rising price of other food. The price of rice was stable because it was controlled by the state. But, as Mr. Zheng told me, the price of vegetables and pork had risen once or twice from 2006 to 2007. When the tea price came down in autumn 2007, other prices didn't. The increasing price of pork made people regret that they had not raised more pigs, but they soon realized that this was impossible: pigs eat corn, but the mountain fields for planting corn had been converted to tea fields. Moreover, they did not have any extra time to work on crops apart from tea.

Some locals began to consider other alternatives. Some quit tea and turned to mining, as Yiwu has lead and zinc resources. Some began planting rubber, even though it was known that rubber should be planted at altitudes lower than 1,000 meters, while Yiwu's average altitude is 1,300 meters. One person who had started planting rubber explained: "In the past five years, Puer tea was of course more profitable. We would stick to it if it didn't meet trouble. After all, it is something we inherited from our ancestors. But you see, rubber has a more stable situation, and it is necessary even during wartime."

Whatever alternatives there were for the locals, rice and other basic subsistence crops like corn and legumes were never forgotten. Many people whom I talked to during my survey wondered whether they should reestablish rice fields, given the uncertain prospects of Puer tea. Several families took such action immediately. The famous saying by Chairman Mao, which many had learned by heart, came into people's minds again: "You will not feel panic if you have rice in your hands" (*Shou zhong you liang, xin li bu huang*).

FOREST TEA OR TERRACE TEA

After the Reform and Opening Up of China, the main impact on Yiwu's tea business turned from the pressure of state policy to the demands of the external market. Since the mid-1990s, Yiwu's Puer tea has been in great demand by outside traders, initially from Taiwan and Hong Kong and later from other areas of urban mainland China. Compared with the slow rise in the tea price from the 1950s to the early 1990s, there has been a rapid upsurge since the mid-1990s, especially after 2003 (fig. 5.5). And as Figure 5.7 shows, since 2004 a price difference has developed between terrace tea and the more expensive forest tea. In the competition between local and outside traders to distinguish "authentic" tea, the tea considered most authentic was forest tea, picked from trees that had not been pollarded, from tea regions with a good, natural forest ecosystem. The most expensive forest tea, which sold for ¥460 per kilogram in the spring of 2007, exemplified this standard. Outside traders and local producers agreed about the basis for this: forest tea is older than terrace tea and has accumulated greater nutritional substance; nonpollarded forest tea is even better, although it grows more slowly than pollarded tea; and forest tea is dispersed in a healthy ecosystem, with sufficient space between trees and good shade from other plants. For all these reasons many people came to believe that drinking forest tea is healthier and has a longer-lasting aftertaste. Terrace tea, by contrast, is younger, planted more densely, and treated regularly with pesticides and

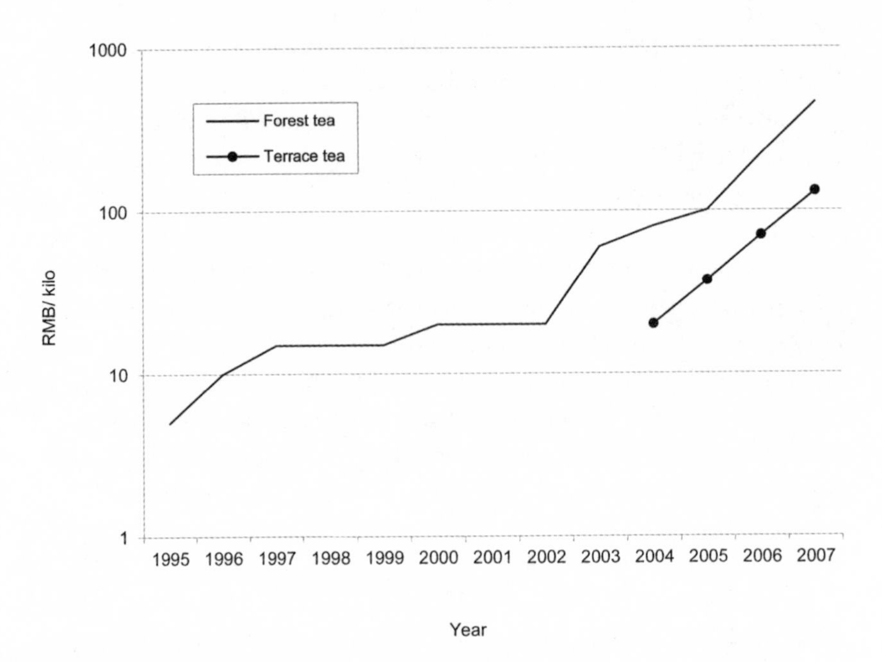

FIG. 5.7 The price of forest and terrace tea in Yiwu, 1995–2007. Synthesized from fieldwork interviews.

chemical fertilizer (although some local people declared that they fertilized their terrace tea only with manure). Of course, picking forest tea from tall and scattered trees requires much more effort than picking terrace tea. Moreover, forest tea's output is far less than that of terrace tea, and its scarcity results in increased price.

But hadn't the locals in Yiwu previously celebrated planting terrace tea, pollarding forest tea, and adopting "scientific" methods of tea cultivation? Yes, but as Mr. Guan said, "Who could predict today's situation? If we had known earlier, we would not have done that." Many locals answered like this, full of remorse. People complained that Zhang Yi should not have helped to popularize terrace tea and the pollarding of forest tea, even though they admitted that he had contributed to the more recent boom in private tea business in Yiwu. The following two cases illustrate how the contrast between past and present standards of tea cultivation generated complex feelings of remorse among many local people.

Of the Six Great Tea Mountains, Yiwu was the one that most enthusiastically adopted terrace tea and pollarding during the early 1980s. And within Yiwu, the subvillages around the center adopted the most modern techniques. This geographical distribution was explained by locals in terms of ethnic difference. Among the Six Great Tea Mountains, Yiwu had the largest population of Han, with central Yiwu as the typical case. Other tea mountains and areas were more ethnically mixed, with Yi, Dai, Yao, Hani, Jinuo, and Bulang. Han people explained that they had always worked harder on tea cultivation than other ethnic groups, and they felt that planting terrace tea and pollarding forest tea was a "scientific" way of improving production. Mr. Zhao, who was still working in the Yiwu government office, once explained to me that Han people were more inclined to obey the rules set by the government. Other ethnic groups were considered backward, lazy, and disobedient.

I wondered whether such "laziness" was rooted in taboos of certain ethnic groups. Some of the original tea planters, such as the Hani and Bulang, traditionally regarded tea trees as sacred plants that should not be cut or damaged (Maule 1991; Shi Junchao 1999; Xu Jianchu 2007). However, my idea was overturned when I visited Gaoshan, a subvillage of Yiwu inhabited by Yi (but whose inhabitants recognized themselves as Xiangtang). People there didn't relate tea trees to any religious worship or taboo, and many attributed their past nonpollarding to their earlier indifference to tea. Gaoshan, 13 kilometers away from central Yiwu, is now famous for its valuable forest teas, which have never been pollarded. While the price of forest tea in central Yiwu was around ¥400 during the spring of 2007, it was ¥460 in Gaoshan. Staying in Gaoshan I saw the locals greet potential clients passionately. And in central Yiwu, I met Gaoshan people carrying baskets of *maocha* for door-to-door retailing. In the opinion of some tea traders in central Yiwu, these Gaoshan people were crafty, often carrying terrace teas from other areas to simulate the forest teas of Gaoshan. Rather than being lazy or backward, they worked as hard as Han people once Puer tea became valuable. In fact, their past "laziness" and "backwardness" had become a great advantage. The people of Gaoshan were not remorseful. Instead, it was the Han people who regretted their once-diligent work.

In the above case, the decision to pollard or not was initially attributed to ethnic difference, but as the situation changed, nonpollarding brought glory to the non-Han ethnic group while pollarding generated remorse for the Han. In the next case, the feeling of remorse was more complex and emotionally aroused.

In central Yiwu, Mr. Hu took me to see his large areas of forest tea, which had been pollarded less seriously. After walking half an hour from his home we came to a grove of tea trees over two meters high. Mr. Hu could not tell the exact age of these tea trees, but he said they had been growing here since he was a child and his mother had come to pick tea from them regularly. He claimed that the tallest and thickest tree was at least five hundred years old, according to a tea expert from Korea. It was the champion among all the tea trees on this mountain, almost four meters high and stretching out on a steep slope beside a valley. I wondered how hard it would be for pickers to work on these old trees.

Mr. Hu said that all these tea trees had been inherited from his ancestor, who had owned most of this small mountain. During the nationalized period, from the 1950s to the late 1970s, the area had belonged to the collective, but Mr. Hu's family had continued working on it. Since the late 1970s, it had been reallocated to his family once again. Many of the trees, even the "champion" tree, had been pollarded following the state's promotion of this technique. But the pollarding was done at a relatively high position on the trunk. Mr. Hu told me that he hesitated to do so at that time, as he thought it was not easy for these tea trees to grow to that height. And the pollarding was not applied to all of the tea trees in the small mountain.

I followed Mr. Hu farther into the forest, where there were more forest tea trees. Although not as tall and thick as the champion, many of these trees were well shaped, with straight trunks and dense leaves. Mr. Hu told me that most of these forest tea trees belonged to his youngest child, Hu Ba. Mr. Hu had eight children, four sons and four daughters. Apart from one son who had died and one daughter who was in Simao, the children lived in central Yiwu and worked on tea fields allocated by Mr. Hu. The three sons used to work in Mengla, but they had all come back to Yiwu to pursue more promising work on tea.

After passing by Hu Ba's forest tea trees, we came to a field of terrace tea, planted in the early 1980s, which belonged to Mr. Hu's third son, Hu San. Although some of the terraces had been transformed from dense into more dispersed plantings, there was no doubt that they would still be called terrace tea. I wondered whether the allocation of tea fields would cause family conflict, since forest tea was much more valuable than terrace tea. And

I wondered if this was proof of the Han custom that the youngest child is favored by the parents. The six children all lived in central Yiwu, as neighbors. Mr. Hu lived with Hu Ba, the youngest one.

I heard more stories when I met the children of Mr. Hu. One story answered my question about family allocation of fields. I was told by Hu Ba that it was an accident that he was allocated forest tea. When Mr. Hu allocated his tea fields to the children, Hu Ba was still working in Mengla. His older brother Hu San was given first choice. At that time, terrace tea was regarded as more "scientific" and easier to manage than forest tea. Hu San therefore chose terrace tea first. When Hu Ba came back, only the forest tea was left for him.

Now Hu Ba, the youngest son of Mr. Hu, was benefiting from working on forest teas. He told me that all of his forest tea was now exclusively sold to a Korean woman under contract. Hu Ba and his wife, Zou, called this woman "Godmother" (*ganma*). She continued her business with the Hu family even after Puer tea encountered difficulties in the domestic market during the summer of 2007. With such a stable client, Hu Ba's family didn't need to worry about finding an outlet for his tea.

Hu Ba's wife told me another story, which was perhaps too sad for Mr. Hu or Hu Ba to recall. Hu Ba's mother used to pick tea from their forest trees. One day she fell from the tallest one, the champion tree that I had seen. It was such a big fall that she died soon after, when Hu Ba was only seven or eight years old. The accident happened around two years before the pollarding movement formally began.

According to the contemporary point of view, it was lucky that some of the forest tea belonging to Mr. Hu's family had escaped pollarding. And these two stories, one about Hu Ba's "fortune" and the other about his mother's misfortune, were interlaced with ambivalent feelings. The death of Hu Ba's mother gave the whole family a good reason to hate forest tea. In Yiwu I had heard other stories of death or injury as a result of picking forest tea. Before the 1990s forest tea was of no value, and people did not work on it as they do today. Some unpredictable and unfortunate events gave them reason to celebrate the pollarding and the new planting of terrace tea. However, that logic was overturned by the Puer tea renaissance in Yiwu in the late 1990s. No one today would doubt that Hu Ba should celebrate his fortunate choice of forest tea. His mother's death, in many Yiwu people's words, was simply a result of carelessness.

Paralleling the change in value of forest and terrace tea material was the transformation in value of fresh and aged Puer tea. In the past, local people threw tea away after a few years. Now, everyone placed value on aged Puer tea rather than raw tea, in the same way that they now appreciated forest tea rather than terrace tea. Both transitions aroused a feeling of remorse, and some locals declared in self-mockery that "people producing Puer tea don't store Puer tea," though paradoxically they kept drinking raw Puer tea due to the limited availability of aged tea as well as the customized palate that had been passed on from the older to the younger generation. Many Yiwu locals told me that these changes could all be attributed to the Taiwanese, who preferred aged tea made from forest tea resources.[6]

Since Yiwu's "discovery" by a group of tea explorers from Taiwan, its private tea business had gradually increased. According to Zhang Yi, from 1995 to 1999 he was the only one making Puer tea in Yiwu, and his clients were predominantly Taiwanese. After 2002, more families emulated him and set up business connections with traders from Taiwan and Hong Kong. By 2004 there were twenty families running private tea businesses (Zhao Rubi 2006: 33), and by 2007 there were fifty families running private tea businesses, whose clients now came from areas in China such as Guangdong, Beijing, and Kunming.

Mr. Zhao, who worked in the government office of Yiwu Township and participated in the reception for the Taiwanese in 1994, said that he had learned tea tasting (*pin cha*) initially from the Taiwanese people: "They taught me to use small tea bowls instead of big ones. They said that tea should be tasted slowly [*man man pin*], rather than simply imbibed for quenching thirst." As he recalled this, Zhao was serving Puer tea in his tea shop using small tea bowls made of glass.

Xu Kun, the head of the Xishuangbanna Supervision Bureau of Technology and Quality, said that many of the recovered techniques for making caked Puer tea were also transmitted by people from Taiwan and Hong Kong, who gave feedback and brought antique Puer tea as a model.

The Yiwu people were mostly full of gratitude for the Taiwanese. They knew that the Taiwanese liked Yiwu's tea, and they had seen that it was the Taiwanese who had initially boosted the tea economy in Yiwu.

During the autumn of 2007 I met Mr. Lü, who led the Taiwanese team to Yiwu in 1994 and had maintained business connections with the locals. He confirmed many of the influences of Taiwanese traders on Yiwu's Puer tea, but he said that the price difference between forest and terrace tea had gone far beyond his expectation. When explaining these issues, Mr. Lü was sitting in the family home of Mr. Zheng. Mr. Lü decided to brew two precious kinds of aged Puer tea to show us what really good Puer tea is. He took out a purple clay ceramic pot (*zishahu*) and small tea bowls from soft cloth bags, which he always carried with him. One of the teas was from the first batch of 7572, made in 1975 at the Menghai Tea Factory, with light artificial fermentation; the other tea was Red Mark (Hong Yin, see fig. 1.3), made in the 1950s also at the Menghai Tea Factory and naturally fermented. When these two kinds of Puer tea were first produced, Yiwu served as the supplier of *maocha* for Menghai. It was acknowledged popularly that the dominant tea leaves in Red Mark were from Yiwu, and the first batch of 7572 was made of leaves of mixed origins. Both teas had been stored in Hong Kong for most of their life and later on traded to Taiwan.

These teas were quite different from the raw Puer tea produced in Yiwu nowadays. The first batch of 7572 was artificially fermented, though lightly, in Menghai, which made it different from the naturally fermented tea. Red Mark, although naturally fermented, was about fifty years old, and therefore different from the raw Puer tea of Yiwu that Mr. Zheng and his son tasted every day in their tea business. The unique herbal medicinal smell that resulted from Hong Kong storage could be detected at once. The two teas also differed from the artificially fermented Puer tea stored in Yunnan that was much more "earthy." According to some people I met, the smell of tea stored in Hong Kong was moldy because of its moist climate, but it was different from the moldy smell that resulted from storage in the moist weather of Yunnan, including that of Yiwu. When tasting Red Mark, I was reminded of soup boiled with rice (*mi tang*), an analogy often applied to good aged Puer tea. Both brews were strong and bright red (*hong nong ming liang*), which was how guidebooks described aged Puer tea (see, for example, Deng Shihai 2004; Zhang Yingpei 2007; Zhou Hongjie 2004), in contrast to the faint yellow raw Puer tea that Mr. Zheng's family and I regularly drank.

There were four tasters: Mr. Lü, Mr. Zheng, Mr. Zheng's son Zheng Da, and me. Mr. Zheng and Zheng Da tasted slowly and carefully. They praised the beautiful color of the tea brew but they didn't give any clear comments on its taste. Mrs. Zheng sat beside us and watched. Like many women in

Yiwu, she supported her family's tea business by doing the cooking and cleaning, but she didn't like drinking tea. She said this was in part because her digestion was not good. In addition, because Mr. Lü served the tea in the evening, she was afraid that drinking it would make it hard for her to sleep. But Mr. Lü encouraged her to have a few small bowls. He told her that good-quality aged Puer tea would benefit her digestion and would not disrupt her sleep. Mrs. Zheng tried some, without significant comment either, and after two small tea bowls she stopped.

Mr. Lü told me that he came to Yiwu for tea almost every year because the most desirable aged Puer teas—such as Tongqing, Songpin, Tongxing, and Tongchang, the so-called first generation of Puer tea—all originated in Yiwu. He had tasted examples of almost all of these teas that were at least sixty or seventy years old, and he hoped the new ones he was collecting in Yiwu would turn out to be as good after several decades of aging. The first batch of 7572 and Red Mark, products of the second generation, also represent the aged appeal of Puer tea, although they are younger. Mr. Lü said that one piece of Red Mark could be sold for ¥85,000 at present, but before the Puer tea market boomed he had sold it to his customers for only several thousand yuan. He generously broke some pieces from his 7572 and Red Mark and left them with Mr. Zheng's family for further tasting.

A few days later I went to Mr. Zheng's house again. The two samples of precious aged tea left by Mr. Lü were untouched. When I asked Mr. Zheng his thoughts on the earlier tasting, he shook his head. He had little interest in aged Puer tea, preferring the raw tea that he produced and tasted every day in Yiwu, which he found full of fresh aroma. He did not care that his clients were counting upon the transformation from the fresh to the aged.

Whether being guided to prefer forest tea, being taught the proper way to prepare Puer tea, or being shown how to delicately drink tea, Yiwu people learned many new rules from the Taiwanese. But when it came to choosing between raw Puer and aged Puer, many Yiwu people still preferred the former, depending on their humble palate shaped by daily life and cultivated by the local water and soil (*shuitu*) of Yiwu.

Zheng Da, Mr. Zheng's son, however, was more interested than his father in exotic things. He tasted the aged tea again several days later with his friends. Like his father, he was used to the fresh tea of Yiwu and didn't praise Mr. Lü's tea, but as time went by he became more appreciative of it. His father, although more loyal to his original palate, recognized that he had to respect the ongoing preference for aged tea if he wanted to make a profit. Though the market recession had brought many uncertainties, he had seen one certainty in the continuing demand of Mr. Lü and others for raw Puer that would one day become aged tea. In the product description he attached to his family-made Puer tea, he wrote clearly that the longer the tea was stored, the higher its value would be.

In July 2009, on a short trip to Taiwan, I drank Puer tea with several young friends who had never been to Yunnan. When they heard that people in production regions like Yiwu previously didn't drink aged Puer tea, they burst out laughing. From their very first experience of Puer tea, they had been informed that Puer tea shouldn't be consumed until it was aged, and they had become accustomed to this law. Their laughter showed that they took aging as a natural and intrinsic feature of Puer tea. They didn't consider the diachronic sequence—that it was fresh tea rather than aged tea that was closer to the original nature of Puer tea, at least for people in the production regions.

CONCLUSION: REFLECTIVE WORRIES

In the *jianghu* world of Puer tea, there are no fixed rules, and all standards are open to the influence of history. Determinants of Puer tea's value have shifted in Yiwu over the past half-century. The stories of remorse and the constant changes in Yiwu echo other stories with similar themes lived by urban traders and consumers, especially those from Yunnan. According to the current standards, the ways Yunnanese treated Puer tea prior its resurgence in the mid-1990s were all "ignorant." In popular books and magazines, and in my own fieldwork, I heard numerous people lament that they should have stored aged Puer tea, made from forest tea leaves, much earlier. This regret encouraged blind buying and storing, often resulting in bankruptcy when the market suddenly receded.

Paralleling urban remorse, this chapter is filled with rural voices. For these people, tea has long been the mainstay of local livelihoods and hence is a source of both anxiety and happiness. Mr. Guan's question about the price of tea reflected the local uncertainty aroused by ongoing events as well as by the past. As in Chinese medicine, the present repletion can be diagnosed in terms of past depletion (Farquhar 2002). Driven by constant changes, there may never be a clear answer about the stable authentic features of

Puer tea. Rather, these features are a response to context and history. The uncertainty of the locals shows their inability to control and adjust the value of tea, yet, while connected to the broader historical context, their worries are not endemic, but contain forces of transformation. External pressures and acquired standards have been used as a weapon by locals in response to the cooling down of the tea business in the autumn of 2007.

CHAPTER 6

Transformed Qualities

> The longer Puer tea is aged, the mellower and smoother its flavor will be. And finally it will attain *hua* [a full transformation]. . . . It will melt in your mouth when you drink it. This is the supreme praise for the best-quality Puer tea.
>
> —Deng Shihai 2004: 49

After I finished my spring fieldwork in Yiwu at the end of May 2007, I stored a box of clothes and other goods at the local guesthouse, so when I came back in mid-September, my stored possessions had spent most of the wet season there.[1] Although I had sealed the box well, a strong smell of mold emerged when I opened it. This gauge of Yiwu's wet season, with rain on most days, caused me to worry about tea that had been stored in Yiwu after processing in the spring.

My worries were partly confirmed over the following days as I drank Puer tea with local families. Some of the tea had the same moldy smell as my stored possessions. The most obvious changes had occurred in the loose-leaf tea. This tea was often kept in a ventilated place, where it was more exposed to the moist air of the wet season and natural fermentation was accelerated. The result was more pleasing if the tea had been put in a mostly sealed room. When tea was stored for around two years under such conditions, the tea brew turned from light yellow to orange (fig. 6.1), with a special plum aroma. But if exposed to moist air for the same period of time, the tea brew turned dark red (fig. 6.2 and back cover of book), with an obvious moldy smell.

The tea traders from urban areas usually appreciated the "plum" rather than the "moldy" smell. But local people, focusing on the color change rather than the smell preference, were happy about the accelerated changes, from a commercial point of view. In their opinion, this transformation proved that Yiwu was a good place to store and age Puer tea, as the tea brew's change in

FIG. 6.1 The darker-colored brew of two-year-old raw Puer tea (left) and the lighter-colored brew of newly produced Puer tea (right), both stored in Yiwu. Photo by the author.

FIG. 6.2 Brew of raw Puer tea stored in ventilated conditions in Yiwu, also around two years old. Photo by the author.

coloration from light yellow to strong, bright red reflected the transformation of high-quality aged Puer teas as described in popular books.

From the traders of Taiwan and the Pearl River Delta, Yiwu people were learning to appreciate aged Puer tea. Although this appreciation may not have been consistent with their original palates, aging had become a handy tool.

As the Puer tea market went into recession, local people applied various strategies to keep their tea businesses afloat, and they looked at the main problems of Puer tea differently from their peers in urban areas. They transformed the newly developed criteria on the value of Puer tea into local practices, thus endowing the characteristics of Puer tea itself with new "qualities."

Practice theory stresses the role of human agency and the importance of interaction between multiple actors in "the dialectic of control" (Giddens 1979: 145)—namely, that control can never be implemented unilaterally. Those being controlled have the potential power to counteract their supervisors with "hidden transcripts" (Scott 1985), such as secret weapons. In this regard, the importance of everyday forms of indirect resistance, contrasted with an earlier focus on peasants' uprisings and revolutions, is underlined and referred to as a "constant struggle" and "subtle sabotage" (Scott 1985: 29, 31). This everyday social action exists with no institutional visibility, no manifestos, and no banner, but the final impacts of the action can be significant:

> They require little or no coordination or planning; they often represent a form of individual self-help; and they typically avoid any direct, symbolic confrontation with authority or with elite norms. To understand these commonplace forms of resistance is to understand what much of the peasantry does "between revolts" to defend its interests as best it can. (Scott 1985: 29)

One aspect of understanding such subtle resistance is to look at how locals explore the margins, which are beyond the definition of existing regulations. That is, by blurring the boundary between their newly explored ground and the already established rules, they can successfully transform the established authenticity into a localized one. Although the attitudes and actions of the peasants in Yiwu could be looked at as a kind of subtle resistance, the word *resistance* is both too rigid, in terms of this particular cultural context, and too dualistic. In the case of Puer tea, there are multiple actors with multiple perspectives. *Transformation* would be a better description. The corresponding word in Chinese is *hua,* which encapsulates neither direct opposition nor absolute weakening but gradual change. The word suggests "melting" and aiming to reach a neutralized state.

The concept of *hua* has long been utilized by Chinese traditional philosophy to refer to ubiquitous transformation, exemplified by the famous discourse of Daoism on the mutual transformation between bad and good fortune: "It is upon bad fortune that good fortune leans, upon good fortune that bad fortune rests" (Lao Tzu 1998: 122–123). In terms of Confucianism, other *hua* forms, such as *wenhua* (cultural transformation) or *jiaohua* (meaning to moralize, domesticate, or civilize), mean to transform something raw or rough, through gradual instruction, education, or domestication, into something civilized and superior.[2]

In the case of Puer tea, *hua* is often used in the term *chenhua* to mean "aging" or "fermentation" in storage, the process by which the astringent feature of raw Puer tea is transformed into a mild, smooth quality. This can occur through long-term natural aging or short-term artificial fermentation.

In the *jianghu* context, *hua* intrinsically indicates the strategies and settlements employed by multiple actors to transform an unsatisfying situation into a comparatively more satisfying one. For instance, some knights-errant depend upon their own martial arts rather than state regulation to help the poor and deal with unfair issues; others withdraw from *jianghu* and become hermits to avoid conflict with the state or other *jianghu* groups; and others find proper ways to resolve disputes between different disciplines and to reach a harmonious relationship. This last approach is called "turning hostility into friendship" (*hua gange wei yubo*).

The above meanings of *hua* shed light on local people's active appropriation of the authenticity of Puer tea. One aspect of the meaning of *hua* is of particular interest—namely, the transformation from something hard and unacceptable to something applicable and digestible.

This transformation and appropriation took place despite the lack of both detailed criteria by which Puer tea's quality could be judged and efficient government supervision. These deficiencies in the definition and regulation of Puer tea may be seen as "margins" that were explored by locals using their transforming agency and self-created solutions and strategies.

Even though the Puer tea market was in recession, the Quality Safety (QS) certification system still applied pressure. The original purpose of QS was to make sure that the tea production process was hygienic, but in practice it resulted in the unavoidable expansion of production spaces and consequent high investment. Under these investment circumstances, locals redefined and transformed the authenticity of Puer tea not only as a form of symbolic representation but also as an important pragmatic tool. Yiwu's tea producers avoided direct resistance to external pressures, but their actions managed worries at a crucial moment, and they empowered locals to achieve a certain degree of freedom.

"WALKING ON TWO LEGS"

In chapter 2, we saw how Mr. He's family tea production was challenged by the new QS standard. During the spring of 2007 he decided to build a new tea factory in cooperation with a capable tea company in Kunming, instead of making changes to his old house.

Mr. He had finished constructing his new tea factory by the autumn of 2007. Had events unfolded according to his original plan, he would have collected more *maocha* and processed it with the greater production capacity of his new factory. But the collapse of the Puer tea market happened just after Mr. He had spent almost ¥600,000 on building his new production space. Furthermore, like many other locals, he had invested in highly sought-after *maocha* during the spring season, when the price was booming.

When I visited Mr. He in the autumn, he expressed his frustration to me and said that it had been an unfortunate year for him. Despite his complaints, he continued to work hard. One day in late September, he got up early in the morning and hurried his son and several workers out to the new factory. It was the day that his new factory would start testing. Only if the test went well could they pass the QS successfully. The new factory, two kilometers away from his house in central Yiwu, was made of reinforced concrete. Its area was more than five times that of the family house. Around an open rectangular courtyard were a number of separate rooms—for storing, sorting, pressing, drying, finishing, and tasting tea—as well as a laboratory, a room for dressing and cleaning, and a factory office.

In one of the two tea-pressing rooms, which measured more than twenty square meters, He San, Mr. He's son, acted as the core link in the production chain. He San was shaping the tea cakes by hand, a difficult procedure that was the most directly determinant of how the final tea cake would look. To his left, He San's fiancée, Little Zhang, was teaching a new employee how to put the correct amount of *maocha* into a cylindrical container and weigh it. She had quit her job in a government office in Jinghong to concentrate on helping with the He family's tea business. I was told that the He family had asked her to help them; if she did not, she would not be accepted as a daughter-in-law. On the right side of He San, two young men were stepping upon the stone press. Here, the traditional way of handcrafting caked Puer tea was continuing.

Next to this was the other processing room, which was not yet in service. A large new machine had been bought and carried in from outside several days earlier. It would be used to press the caked tea mechanically, as an alternative to the stone press, but Mr. He and his family were still waiting to learn how to use it. This machine was a necessary part of the QS, and it would be used mainly when producing gift tea (*lipin cha*), tea intended to be used at special events for gift giving. Gift tea was assumed by some people, though not all, to be made of poorer-quality tea leaves, but demand for it could be high during special seasons such as the Chinese New Year and the Mid-Autumn Festival, when Puer tea was given as a gift in urban regions. The He family's production choice hinted that better-quality tea material should be paired with the traditional method—the stone press—while poorer-quality material was left to the machine, which could increase the production output. But the machine modernized the new factory, in contrast to the stone press in the nearby room, and to all the other processing equipment in the family house in central Yiwu.

Mr. He supervised all the work and dealt with matters both trivial and important. He had complex and ambivalent feelings about the modern space. He generally didn't like it, and he complained that it forced him to spend a large amount of money, but at the same time he was very proud of his new construction. He took a group of guests from Beijing on a tour of the factory and hinted to them that this was the best tea production unit in Yiwu. The guests were curious about the "tea laboratory," where there were professional evaluation teacups, test tubes, and an accurate scale, which were usually found only in chemical laboratories or at special meetings for tea evaluation. Mr. He hinted that this equipment might not be used in practice but was required by QS.

The new factory also included a kitchen and lodging for the employees, but due to some unfinished work, for the first few days of operation the workers had to go back and forth to the old family house in central Yiwu. At home, Mrs. He and her daughter, Hongping, cooked lunch and dinner for them. Mrs. He was worried because she needed to look after the family house, but in the future she might have to go to the new factory to cook, to avoid hiring another cook. Hongping ran a tea shop in Kunming and came back home temporarily to help with the housework as well as factory management. In her words, the initial test for the new factory was "a big thing" and these early operating days were "a crucial moment" for the He family. As I learned later, the family received some orders in the autumn, though general business was still slow. They worked hard to recover some losses and, of course, to learn to use the new factory. The actual inspection and evaluation would take place soon.

Staying with them throughout the day at both the factory and at home, I could feel the tension that the tea business brought upon the whole family. Even during lunch and dinner, Mr. He and Hongping reminded the new employees about procedures in the factory. After lunch and a short break, Mr. He urged He San and the other workers to go to the factory again. He San drove the workers back to the factory first. Mr. He then followed on foot after finishing other things at home. It was not until after dinner that he could sit down at home and enjoy smoking his bamboo pipe in his quadrangle courtyard, with its flagstone patio and wooden structures: pillars, windows, and stairways. This courtyard and another connected to it had been the area for processing tea. The kitchen, bathroom, living room, dining room, and bedrooms were all nearby on the ground floor. In Mr. He's words, "All things were arranged simply, handily, and traditionally." The upper level of the house had been used for packing and storing tea. The upstairs windows provided a view of the tea fields on the opposite mountain. While smoking his pipe, Mr. He explained to me how he had realized the important value of this old space:

> There have been many times when travelers entered my house without invitation. And usually, in the end, we did tea business with them. I have been wondering why this could happen. As you know, there are many families selling tea in Yiwu, but in some people's eyes I am more successful. I think it is largely due to my traditional house. If I leave my door open, people passing by notice that there is something special inside. They are curious about my house and, of course, about how we make Puer tea in the handcrafted way. Yes, the new factory is excellent, but I know that few visitors will go there. Therefore, I insist that at least one of my family members stay behind to look after the old house.

Mr. He's family house was more than seventy years old, and it had been used for processing Puer tea for three years before the new factory was built. It was on the government list of Yiwu's protected architecture. Realizing the value of his house, Mr. He decided that he would keep both spaces for tea processing. The new factory would acquire QS certification soon, and in Mr. He's logic, this certification could be extended to cover his house, at least for demonstrating the "traditional" way of processing tea. QS regulations strictly defined how to format the modern processing space in order to have clean production and safe consumption, but it didn't declare that the traditional space must be abandoned. By making use of this regulatory loophole, Mr. He sought to maintain the modern and the traditional at the same time. In his words, he did this to "walk on two legs." However hard it was for him to go back and forth between the two places, he persisted in doing so every day. Later, he made his multiple journeys easier by learning how to drive a car, despite the fact that he was over sixty years old. Moving with modern transport rather than on foot, the distance between the modern and the traditional was seemingly shortened. And using a flexible logic, he successfully integrated his modern and traditional production spaces. As a result, the Puer tea processed by his family, whether in the modern factory or in the old family house, would be marked with both QS certification on the wrapping paper and the indispensable words "handmade traditionally by a family" (*chuantong jiating shougong zhizuo*).

"WAIT AND SEE" AND THE "BLANK VERSION"

Mr. He's new tea factory passed QS in October 2007. At that time, according to official data, around fifty Yiwu families had officially achieved QS for their fine processing. But there were actually far more than fifty families involved in fine processing, perhaps as many as eighty. Of the extra units, some were still struggling to construct a new production space in order to obtain QS certification, while others never planned to do so. According to those in the latter camp, QS was not a serious issue and they could still continue fine processing in their own way. Hu Ba, the tea producer who had a stable business contract with a Korean godmother, was one such case.

Hu Ba didn't have an attractive old house like Mr. He, but clients still came to him because of his excellent and famous forest tea resources. He processed his own tea in his family house, for both rough and fine processing. He didn't need to worry about having an outlet for his tea, even after the problems in the market, because of his relationship with the Korean godmother. He said he had no ambition to run a large-scale tea business and would base his business on his present tea resources and sell them to stable clients. According to him and others in similar situations, QS meant making a large investment, which was not worth doing for small-scale production. The strategy Hu Ba adopted was simple and practical: he worked only when the officials were not coming for inspection. Such unregulated moments, according to his experience, were abundant. He felt more confident when autumn came, as QS officials seldom visited after the Puer tea recession. He congratulated himself when he witnessed the losses incurred by families who had been working hard on expanding their production facilities. Most importantly, he had stable clients who continued to buy his tea products even though he lacked a QS certificate. The Korean godmother was his biggest client. Rather than making changes in fine processing to achieve QS, which was only a provincial standard, Hu Ba's family made changes in rough processing as requested by this stable client from overseas. His wife, Zou, told me about the Korean godmother's requests:

> She said that QS doesn't matter to her, but she reiterated that we must avoid using any pesticides or fertilizer on the tea plant. She declared that she would have a chemical examination done on the tea that we sold to her. If any index was above the standard mark, she would reject our goods, because she would have trouble exporting them to Korea. And of course she asked us to harvest, process, and pack forest tea and terrace tea separately. She was a skilled tea taster, you know. She also asked us to dry the tea leaves in the sun rather than baking them with fire or a machine. So you see, during continuously rainy days, I would give up harvesting and just let the tea grow. Moreover, she asked us not to trim the tea trees in winter, lest the plants bloom too fast in the coming spring, which would reduce their quality.

Hu Ba's family accepted all of these rules even though they were much stricter than the government's suggested standards for rough processing. Throughout 2007, most of their tea products were sold to the Korean trader at market price. Encouraged by this success, Hu Ba began increasing tea cultivation, as the Korean woman said that newly planted terrace tea would be acceptable as long as pesticides and fertilizer were not used. This meant that his production scale could be doubled in the near future. When the topic touched upon QS, Hu Ba's attitude was "wait and see." He wasn't sure about his future production mode but he could proudly declare that he was producing authentic Yiwu tea.

One day I visited one of Hu Ba's relatives, a woman who lived not far from his family. She also lacked QS certification, but Hu Ba's family brand was more famous. Nevertheless, the woman insisted that her tea was as good as Hu Ba's. She showed me one cake of her own tea wrapped with simple white paper, without any characters or images, which many locals called "blank version" (*bai ban*) (fig. 6.3). The woman explained that she didn't want to spend extra money printing colorful wrapping paper and stressed that this generic Puer tea without any QS mark was specifically requested by some special clients because it represented authentic tea produced by a small-scale family business. Tea products with QS, on the other hand, were seen by some people as ugly and inauthentic, representing larger-scale production that focused on quantity rather than quality. I also understood that some outside traders specifically ordered these blank versions so they could rewrap them with their own brand's paper.

This woman's family was one of the twenty-three families in central

FIG. 6.3 The "blank version" of Puer tea. Photo by the author.

Yiwu that I surveyed in October 2007. My original intent was to acquire quantitative data to measure the change in cultivation after the blossoming of Puer tea. After answering my survey questions, many of my interviewees raised interesting issues beyond my questions. Quite often they asked for my opinion on the future prospects of Puer tea and on what was happening in the urban market. And almost everyone actively and sharply criticized the heavy intrusion of tea imported from outside Yiwu, which was taken to be the root cause of local worries. According to them, the collapse of the whole Puer tea market was a result of concern about inauthentic tea, and because of this blemished fame, fewer visitors would come to Yiwu or buy Puer tea. According to local analysis, the incursion of non-Yiwu tea leaves was related to the operation of QS. Their logic went like this: QS demanded a high investment to enlarge the fine processing area. Once families had made this investment, they had to collect enough *maocha* for their enlarged production scale and, of course, to recoup the QS expenses associated with higher-quantity production. However, as everyone knew, the limited tea resources in Yiwu could not support the increase in fine processing or in the number of tea traders. As a result, one solution was to import cheaper tea resources but to mark the finished product as "authentic Yiwu Puer tea," along with a QS certificate on their paper wrapping.

Caught between government regulations and specific market demands, the local producers had to make a careful choice. The QS standard required "addition": adding facilities and enlarging the production scale. By contrast, some traders yearned for "subtraction": reducing undesirable work on tea cultivation, keeping small rather than large processing units, and sometimes even leaving the wrapping "blank." People like Hu Ba and his relative didn't follow government regulations, nor did they oppose them openly. They believed that "addition," the larger-scale production with QS approval, actually produced inauthentic Puer tea. At the same time, in tacit agreement with their clients, they legitimized their work without QS—that is, "subtraction"—as the authentic way.

The QS requirements could also be read as "tunnel vision" (Scott 1998: 13)—namely a radical simplification managed by the government to help impose taxation and to mold monoculture. In the case of Yiwu Puer tea it also stood for standardization and mechanization. This simplification, then, was deconstructed by locals like Hu Ba and his relatives with flexible strategies. The standardization came up against the attitude of "wait and see," and the mechanization was counteracted by more complex hand work in tea refining, as requested by clients who sought specialty goods. So by blurring the boundary between regulation and practice, they readjusted the simplification and turned it back into complexity. It is not only that "what was simplifying to an official was mystifying to most cultivators" (Scott 1998: 48), but also that what was made complex by the locals was mystifying to the officials—or, at least, hard for the officials to judge or regulate.

TU CHA: INDIGENOUS TEA

Gao Fachang, a math teacher at the local Yiwu middle school, worried about something different. To him, the most important issue was the damaged ecosystem of the tea mountains.

Before I visited Gao in person, I had seen copies of a map of the Six Great Tea Mountains that he had drawn hanging in many local houses. Local people loved it, and it was nearly indispensable for families running Puer tea businesses.

But along with praise for Gao, people in Yiwu described him as a meddling odd man who was overly critical and behaved eccentrically.

When I met him, I was more shocked by the price of his map than by his short stature, which contrasted with his name (in Chinese, *gao* means tall). The map cost ¥40, enough to buy almost four good-quality color maps in urban bookstores. In addition to mapping the location of major tea resources, Gao's map showed past and present transportation routes, the location of historical relics such as temples and stone tablets, and longitudes and latitudes. This was not just a schematic illustration, as appeared in other popular books about Puer tea, but an "integrative map," as Gao called it. He had done all the surveying and mapping himself. Because he lacked advanced surveying instruments, he made measurements on foot and relied on local experience as well as conversion and calculation based on previous maps. The map took him four years to create, and he published it privately, because he could not meet the difficult conditions set down by professional publishing units. "That's why it is expensive," Gao said, "and why you must buy it directly from me. You can't get it at any bookstores.".

While looking at the map, he gave a critical territorial scoping on the production places for authentic Yiwu Puer tea: "Not just any tea growing within this geographical boundary can be defined as authentic Yiwu tea. Tea trees planted mixed with rubber trees and below 1,000 meters altitude, I think, should be excluded." He referred particularly to a place called Nametian, which was a twenty-minute drive from central Yiwu. In the early 1980s, when terrace tea was first cultivated, nine tea teams (*chadui*)—actually nine large terrace tea fields—were established in Yiwu, with eight located around Nametian and one in central Yiwu. The area in Nametian was later used for planting rubber trees, which were mixed in with the tea trees (fig. 6.4). Gao said,

> You know, certain insect pests are fond of rubber, and pesticide must be applied. You can imagine how the tea planted along with the rubber would suffer. How could this tea share the general glory of the good tea of the Six Great Tea Mountains? Good-quality tea should be on high mountains above 1,000 meters, but Nametian's tea and rubber fields are only at 700 or 800 meters. In my opinion, among the nine tea teams, only the terrace tea field in central Yiwu should be regarded as authentic.

Gao then told me two stories about ecosystem damage, stories that may have contributed to the image of him as an "odd meddler." One time, he

FIG. 6.4 Tea (the lower bush form) planted along with rubber. Photo by the author.

witnessed someone cutting down two old tea trees. He reported the incident at once to the local Forestry Bureau but did not receive any response, so he called the higher level of the bureau. Still nothing was done. He was angry and warned the forestry official that he would take matters into his own hands, implying that he might kill the tree-feller. Finally, the officials were forced to come. Another time, he wrote a letter to the central government in Beijing informing them that rubber fields were being increasingly planted in Xishuangbanna and that several local government officials were illegally involved. His letter received a positive response, which made him famous, but the local situation did not change much.

Gao's negative view of rubber was based upon his own experience. Once, he took back a small portion of soil from a rubber plantation. When the wet season came, all the soil in his field sprouted grass except that from the rubber plantation. He provided a comparison between rubber trees and tea trees:

> Tea can coexist with other plants peacefully. But rubber is like a pump. It's hard to find any other plants in the rubber forest. The earth used for rubber may become dry and eventually useless . . . The history of rubber planting in

> Xishuangbanna is less than one hundred years old. We still don't know what its impact will be, just as we still lack knowledge on aged Puer tea.

Gao said the size of the tea area, especially that of forest tea, was hard to estimate. But since rubber was always planted in a large area, based upon his own measurements on foot, he estimated that the present area of rubber in the Six Great Tea Mountains was no less than 500,000 acres. "Worst of all," he said, "many local people planted rubber trees after the Puer tea market went into recession. And they don't even care that rubber should be planted below 1,000 meters altitude. This will have a more negative impact on the tea trees." To him, the downturn of the Puer tea trade was normal and perhaps temporary. Only if the tea trees were protected well would the Puer tea business bloom again. But once the ecosystem was destroyed, the prospects for Puer tea would be bleak. He did not think that people should refrain from planting rubber altogether, but he felt strongly that the rubber area was expanding too quickly.

Some locals sneered at Gao's concern as a groundless worry. Others thought his behavior was too radical. But quite a few people acknowledged the validity of his argument and appreciated his stories. Some urban visitors also echoed his concern and helped him publish articles anonymously in newspapers and on websites. Gao also wrote a book about the history of Puer tea in the Six Great Tea Mountains. As with his map, however, he had difficulty publishing the book. The main problem, according to the publisher in Kunming, was that Gao was too critical of other tea experts and celebrities who had written about Puer tea and whom the publisher did not want to offend. Gao said he didn't understand why he could not criticize these "experts," and asked, "Shouldn't the authentic history of Puer tea come out through debates?" Gao was proud of his own writing and regarded his account as the most indigenous and therefore authentic, as most other accounts of the Six Great Tea Mountains had been written by outside traders, urban writers, and government officials. But a year after submitting his manuscript, Gao's publisher was still asking for revisions. He was being asked not only to limit his critiques but also to revise his writing style and adopt a structure more suitable to a formal publication. Gao could not help but feel that the more intrepidly he acted, the more his feet were bound.

Despite his many concerns, Gao said he could still sleep well at night, as long as he drank good Puer tea (mostly his own) during the day. Gao

FIG. 6.5 The tea wrapping designed by Gao, reading (from inner to outer circle):
1. Earthy tea (or "indigenous tea").
2. Xiangchang (a brand of tea) from Xikong (the hometown of Gao in Xiangming township). Quality with good faith. Tender buds from aged tea trees. The older the better.
3. Purely natural and organic Puer tea cake, produced with tea material from Xiangming that includes four of the Six Great Tea Mountains, the original tea-growing place in Yunnan. Photo by the author.

ran a small-scale tea business in Yiwu, using the processing unit of a close friend for fine processing. He collected *maocha* by himself and chose the leaves he liked for his caked tea. He supervised the entire pressing process and designed the paper that was used to wrap the final tea product, the main color of which was earthy red. Characters and patterns were placed in concentric circles. Most remarkable were the two characters in the central circle: *tu cha,* which could be read as "earthy tea" if literally translated, or "indigenous tea" by extension. By using the word *tu* (earthy), he said he wanted to indicate that each piece of Puer tea was an indigenous product (*tu te chan*)—authentic but also common, not magic, as Puer tea was depicted by some. Text in the outer circle further clarified the meaning of *tu:* the origin of the tea leaves was from Xiangming, Yiwu's neighboring township, which included four of the Six Great Tea Mountains (Manzhuan, Mangzhi, Gedeng, Youle). Gao was born in Xiangming, and he emphasized to me that he identified his ethnicity as "indigenous" (*ben ren*), rather than Yi, which is how local minorities are identified by the government. He didn't collect *maocha* in Yiwu much, as in his eyes Yiwu's ecosystem had been destroyed. The front of the tea wrapping did not have space for the QS mark, although the production unit of Gao's friend actually had achieved this qualification. Like some other locals, Gao saw his small-scale production as authentic and disregarded QS. But unlike Hu Ba's relative, who ignored QS by presenting a "blank version," Gao made a "full version" by filling all the corners of the packaging to describe his indigenous ideas. Despite confronting so much trouble and so many restrictions, his self-designed product became the item with which Gao could freely exercise his ideas about what constituted authentic Puer tea. He reminded his customers that the tea was "purely natural and organic" and would benefit from aging (fig. 6.5).

CONCLUSION: MULTIPLE TRANSFORMATIONS

These case studies show how local people redefined the authenticity of Puer tea flexibly and pragmatically. Rather than directly resisting the government's rules, they deployed strategies focused on exploring the margins of regulation and finding spaces where local voices could be legitimized once again. Producers actively appropriated the authenticity of Yiwu's Puer tea no matter how government regulation or market values changed. Additionally, outside forces—such as the Taiwanese influence, which had a major impact on transforming Yiwu's nature and culture over the past ten years—were recontextualized and transformed into local pragmatic strategies. Although the difficulties couldn't be solved completely, these multiple transformations did address some local concerns. As in the Daoist saying "It's upon bad fortune that good fortune leans," these Puer tea producers relied on self-sufficiency to gradually transform a difficult situation into a more pleasing one.

Moreover, as part of the dynamics of transformation, one transformed quality is always destined for further transformation. "The older the better," the strategy that celebrated long storage of tea and partially rationalized slow turnover of inventory, was soon challenged. As a moldy smell arose from storage facilities, it was admitted that "the older the worse" might also be true if the aging process was not managed carefully. As a result, "retouching," new transformation, and redefining began, stimulated by both external preferences and indigenous realization.

冬 WINTER 藏

Puer tea stored in a Yiwu family house.
Photo by the author.

CHAPTER 7

Tea Tasting and Counter–Tea Tasting

The other world isn't so pure; the other law isn't so perfect, either. A real Chinese knight-errant needs to retreat not only from the court but also from the *jianghu*, like the characters in [Jin Yong's novel] *Beyond the Rivers and Lakes: The Smiling, Proud Wanderer.*

—Chen Pingyuan 1997: 176

More than fifty people attended a tea-tasting event at a Kunming teahouse in November 2007. The purpose of the event was to discuss whether older tea is better and what kind of storage produces good taste. Sanzui, one of the most influential tea websites in China,[1] organized this meeting. At the time, the Puer tea market was in recession. Disputes about whether the quality of Puer tea depends on aging had emerged frequently on the Sanzui website, turning the virtual space into a battlefield. Mr. Yan, the moderator of the Sanzui web column on Puer tea, had announced, "Rather than disputing nonsense all day long, we should talk face to face by sharing and discerning real aged Puer tea." So, gathering in a public tea-tasting space, people began to explore issues relevant to the time and space of tea storage.

Six kinds of Puer tea, aged from six to nineteen years, were served at the event. Three were aged raw and three were artificially fermented. They were selected from many samples contributed by friends of the website from places including Yunnan, Guangdong, Hong Kong, and Taiwan. Any member of Sanzui could attend the tea tasting, providing that they promised to write a comment of at least a hundred words on the website afterward. Being restricted by distance, most participants were from Yunnan, especially Kunming. Other members of Sanzui in places such as Guangdong, Hong Kong, and Taiwan could share in the event only via the web posts, which provided live coverage of the tea gathering. Among the participants, experience or knowledge regarding Puer tea drinking was uneven. Some were experts, while others were just beginners. Their social backgrounds were mostly unknown. From their taste preferences, exemplified by the tea comments they gave later, it was obvious that many sought to improve their economic and cultural status by investing in and appreciating Puer tea.

I had begun participating in this online forum after getting to know several key organizers of Sanzui, such as Mr. Yan, in Yiwu during the spring of 2007. At that time they were collecting information on Puer tea and searching for "authentic" tea from the tea mountains. In the Sanzui circle, those who had been to the tea mountains were acknowledged as having a greater right to speak. A tea-tasting meeting, to some ordinary enthusiasts of Puer tea, was a good chance to meet tea experts and taste the tea recommended by them.

The organizers of this tea meeting, like Chinese knights-errant wandering in *jianghu,* had attempted to solve all disputes on Puer tea through organizing a singular event, counting on personal tasting skills and trying to isolate tea tasting from other influences. This attempt failed, because one distinct interest met, contested, and interacted with another, and counterforces circulated between the different actors. Tasting judgment was inevitably affected by the atmosphere and social interaction with other actors.

Anthropologists and sociologists of food have argued that taste is influenced by many exterior factors beyond innate palate preference (see Messer 1984; Mintz 1985; Sutton 2001; Strasser 2003; Lien and Nerlich 2004). Case studies also show that taste judgment often results from the mixed standards of internal preferences and external symbolic values; when the taster's prior value standard conforms with the symbolic meanings attached to the food, it tastes good, and vice versa (Allen, Gupta, and Monnier 2008). Food becomes something not only to eat but also to think about. Generally, a person's distinct way of consuming food actually reflects his or her distinct self-representation (Ohnuki-Tierney 1990; Lupton 1996; Mintz 1996; Caplan 1997; Miller 1997; Gabaccia 1998; Counihan 1999; Counihan and Van Esterik 2008).

These different self-representations are often strongly influenced by different educational backgrounds and social origins; each group has its unique social space and "cultural capital" (Bourdieu 1984). This was strikingly true for the Puer tea event, in which tea taste appreciations differentiated one person from another and acted as signs of cultural status. For social distinction, previous studies have emphasized the gap and distance between different groups of people (Bourdieu 1984). The Puer tea-tasting event, however, provides an opportunity to look at distinction from the point of view of interaction rather than only distance. When all actors' distinct self-representations are juxtaposed, not only their difference, separation, and distance but also the way that they relate, interact, and deal with one another become apparent. That is, the way an individual constructs himself or herself is also the way that he or she joins in constructing social relations; and the standard that one uses for self-representation is based upon referring to others.[2] In the tea-tasting event, each group tried to show a unique way of identifying Puer tea by avoiding others' influence. Some participants even opted to retreat rather than directly compete, making the *jianghu* battle relatively still and secretive rather than open and violent. Nevertheless, the attitudes participants took to separate themselves from others clearly expressed the ways that they countered, related to, and viewed one another.

In the *jianghu* of the tea mountains, the anxiety about finding authentic tea goes beyond technique and is transformed into anxiety about the struggles of human negotiations. Likewise, in the *jianghu* of the teahouse, the anxiety of identifying where a particular tea has been stored is transformed into anxiety about interactional social spaces.

AGAINST HUYOU

Throughout 2007, in Kunming, numerous tea-tasting events (*chahui,* literally "tea meeting," or *dou cha,* "tea competition" or "tea game") were organized by both government and nongovernment tea associations (fig. 7.1).

These events emerged within particular contexts. First, they became more common at a moment when Puer tea was experiencing extreme popularity. Traders used them to let more people know about their tea products and collections; mass media outlets that started columns on Puer tea were eager to collect material for publication; individual tea enthusiasts were keen to meet and test their knowledge and tasting ability. Puer tea became something not only good to taste but also important to know about. After recession struck in the summer of 2007, the passion for holding tea-tasting events was diminished, but they still continued through the winter. The uncertain prospect of Puer tea made people worry, and they were eager to discuss their concerns with others.[3]

Second, holding tea-tasting events was influenced by economic develop-

FIG. 7.1 *Chahui* held in a tea restaurant in Kunming in 2007. Professional tea infusers wore ethnic costumes (top), and participants answered questions on a piece of paper after the tasting (bottom). Photo by the author.

ment in China. Antique things were despised during the planned economy period (1949–1978) and especially during the Cultural Revolution (1966–1976). After the Reform and Opening Up (the late 1970s and the early 1980s), the trend gradually reversed. Spurred by the economic upsurge since 1992, China had been modernizing, and people were celebrating new lifestyles. But at the same time, the value of the ancient had been rediscovered. Tea, a national drink that had been ignored for fifty years, became a focus of attention once again. The popular tea-tasting event in Kunming was a kind of imitation and recovery of the ancient tea competitions and tea gatherings, popular tea events in the Song (967–1279) and Ming (1368–1644) dynasties.

Third, each organizer wanted to present a special way of tasting Puer tea, different from the others. Therefore, the rivalry existed not only between participants in one tea-tasting event but also between different tasting events. The organizers of the Sanzui tea tasting were potentially competing against several "others." They were against those traders who didn't really love or know about Puer tea but were just good at advertising and cheating. They declared that they would provide the quintessential Puer tea at their event to prove their supreme connoisseurship. Though they didn't intend to oppose the government regulations on tea, many of them looked down on those regulations, such as the updated definition of Puer tea set down by the government of Yunnan. The organizers and participants said that they could easily find mistakes in official definitions, and some didn't care how Puer tea was defined at all. They also didn't trust academic tea experts, who, in their eyes, merely worked in the laboratory, seldom stayed long in the tea mountains, and didn't have enough experience tasting real aged Puer tea. The Sanzui event was particularly opposed to those who spread gossip that downplayed Puer tea's merits after the market crisis. By holding a large tea-tasting event, the organizers wanted to give participants a chance to experience the value of well-aged Puer tea.

In an attempt to counteract the above agents, the organizers of the Sanzui event emphasized one thing: each individual's own unique taste ability. A famous post on the Sanzui website in 2007 was titled "To immunize against *huyou,* be loyal to your own senses" (Shengse Chama 2007). Originally *huyou* was commonly used in China's northeastern dialect to refer to irrelevant, exaggerated, or nonsense words (*baihua*), which often reduce the judgment of the audience or create vagueness. Ever since the term was used by Zhao Benshan, a famous comedian from northeast China, in his performance at the 2001 CCTV's Spring Festival Gala, *huyou* has been popularly used to refer to cheating, swaying, hyping, wheedling, or agitating someone with fictitious content. Compared with the serious crime of fraud, *huyou* implies a milder sense of cheating, and it is often used in a joking or sarcastic context.[4] Since Puer tea's rise in popularity, *huyou* has also been widely used to suggest strategically swaying, directing, cultivating, or infecting others in order to promote tea sales or personal fame. As a result of *huyou* the flavor of tea becomes an acquired taste, rather than one based on innate feeling and judgment. This Sanzui post revealed the fact that Puer tea had been largely shaped by many "packaged" values—namely, all sorts of exaggerated statements and mistaken information. The web post encouraged people to evaluate Puer tea based on their innate preferences rather than other external factors. That is, faced with the risky *jianghu* of Puer tea, one must depend upon one's own tasting skill in order to discover authentic tea. This tasting skill is simple, as it avoids exterior interference, though it is also complicated, because it relies on innate talents as well as diligent practice.

One month before the Sanzui event, I was at another tea-tasting event organized by a local newspaper. Mr. Yan, the organizer of the Sanzui event, was there too. Over the course of about two hours we were served eleven kinds of Puer tea, most of them quite young. Due to the volume of tea being served in a limited time, each tea could be tasted for only two or three "runs."[5] According to Mr. Yan, this abbreviated tasting did not show how long the tea could stand up to being infused. The tea serving was at times disorderly and confused. After tasting, the participants were asked to guess the age and production origin of each tea. Those with the highest scores received rewards, and those with the lowest were required to buy something. Mr. Yan was obviously unsatisfied with the tea, the serving, and the regulations. It was at this event that he told me about the forthcoming Sanzui event, which, he said, would allow people to fully enjoy drinking Puer tea. He later put his promise into practice. The Sanzui event lasted longer, with each tea being fully infused; it set different competitive regulations for the participants; it provided more types of aged Puer tea that had been infused using superior techniques; it created a distinguished atmosphere for the event; and, most importantly, it asked participants to be loyal to their own senses and to express their true opinions about the teas' flavors.

ANCIENT PLUS MODERN

The Sanzui event was held at the teahouse of Mrs. Fan, who was well known in the Sanzui community as an expert tea infuser. The teahouse was located beside a lake in a quiet residential district, in keeping with the old custom that a teahouse should be remote from noise and distraction. I went there after a full lunch, knowing that I would not be able to bear drinking so much tea on an empty stomach. When I arrived, the ground floor of the teahouse was crowded with guests, all watching Mr. Yan, the organizer, allocate six kinds of Puer tea. Chinese zither music was playing in the background, providing a melodious and relaxed tone to accompany the tea drinking, although the effect was negated by the large crowd.

Although it had modern elements and materials, such as Western-style sofas and lights, the teahouse largely imitated an ancient Chinese style of décor, with elements like a round latticework wooden screen, Ming-style chairs and tables, and calligraphy and traditional Chinese flower and bird paintings on the walls. In many corners stood ancient-style vases and jars, and delicate tea sets lined the shelves and tables. All of these features contributed to give the teahouse an "antique feel" (Clunas 1991: 81). The physical presence of Puer tea strengthened this point. It was used as the most remarkable antique decoration in the teahouse, both in the form of actual aged Puer tea in various shapes and as illustrations of past Puer tea brands.

The tea-tasting event aimed to create an ancient aura. This included the decoration and the atmosphere of the teahouse, the method by which tea was infused, and the way the event itself was carried out. Great attention was paid to the serving of the tea, displaying a practice that takes great effort, as summed up in the standard Chinese term *gongfu*. It is said that only through very careful and highly skilled infusing can the intrinsic quality of tea be fully presented, enabling participants to judge the storage feature properly. All the tea at the event was infused in Chinese purple clay teapots, as this type of pot is acknowledged as the best vessel in which to make tea. Each type of tea was infused separately, as clay pots easily absorb smell. Thus, the pot for oolong tea and red tea couldn't be the same, and raw Puer and artificially fermented Puer were kept apart. The tea brewed in each pot was poured into a serving pitcher and then into individual tea bowls. Pouring the tea from the pot into the serving pitcher first assured that every guest's cup of tea would be of the same strength and consistency, symbolizing that each person around the tea table was equal. Both the serving pitcher and the individual tea bowls at the Sanzui event were made of glass so that the color of the Puer tea brew could be better appreciated. This helped tasters to judge the age of the tea. Next came a stricter principle that had been ignored by other tea tastings, but was followed closely at the Sanzui event: different infusing techniques were applied to different teas. This included varying the temperature of the water, the height and direction from which water was poured into the pot, and the steeping time for each run. These infusing techniques could not rival the Japanese tea ceremony in their attention to detail, as their purpose was more taste sensation rather than spiritual inspiration. These careful techniques, which embodied the craft of tea infusing (*gongfu cha fa*), originated in the Chaoshan area of southeast China, where they were developed mainly for oolong tea, and they had now been borrowed for many other types of tea, including Puer. The individual craftsmanship of the tea infuser ultimately decides how these details are carried out. At the Sanzui tea tasting, three expert tea brewers, all of whom had obtained the certificate of craftsmanship in tea infusing (*cha yi shi*), were present, including Mrs. Fan. Each attended to one of the three tea tables. Each kind of Puer tea was brewed for at least five runs, unlike at the previous events that Mr. Yan had attended and been dissatisfied with. Furthermore, after the five formal runs the tea was transferred into a bigger pot, which was then boiled for ten minutes and served, in imitation of the ancient way of boiling tea. According to the organizer, this final boiling released the remaining substances in the tea so that the value of the precious aged Puer tea could be fully appreciated.

The ancient atmosphere at the Sanzui event was documented by modern technology. Next to the registration desk in the teahouse was a computer with an Internet connection. Three or four people who were involved in organizing the tea tasting worked on the computer, writing posts on the Sanzui website that reported on what was happening in the teahouse (Sanzui 2007). Any participant who was a member of Sanzui could do this, too, though most were fully engaged in tea tasting. Other members of Sanzui who could not attend the party responded online to present their exclamations, jokes, and questions.

Those doing the web coverage made sure to go back to the tea table to fill their teacups at the right moment and to collect more material for their posts. Taking photos was a key task. Many participants had digital cameras and took several photos of each tea sample: in its pressed shape, when it was infused with good color, and when it was tasted. Several photographers from the mass media were present, and they later reported on the event in publications such as *Puer Jianghu* (magazine), *Pu-erh* (magazine), and *Kunming Daily* (newspaper). At certain points, so many cameras flashed that it seemed that the photographers were recording a "historical moment." After all, the six kinds of tea were precious and could not be obtained easily. I had brought a video camera with me, and at the event I met another person who also had one. He had been invited by the teahouse owner to record the entire event. The Puer tea served at the Sanzui event was recorded and digitized by both still and video cameras. It became modern.

SILENT CONTEST

The Sanzui tasting was a large event, with guests coming and going. The tea tasters were separated into three groups in the teahouse, one on the ground floor and the other two on the first floor. Each group sat around a tea table and had its own tea preparer, but the same kind of tea was prepared and tasted simultaneously at all three tables. The ages of the first five teas served were clearly marked at each tables. The sixth tea's age was unknown, even to its contributor, but there was no doubt that it was aged. The six types of tea were infused one by one, from the youngest to the oldest, with raw tea alternated with artificially fermented tea. Many previous tea-tasting events had asked participants to guess the age and the production origin of the teas, but the Sanzui event adopted a different procedure, asking the participants to judge the corresponding storage details after tasting. The Sanzui event avoided the examination-like style of other events, which asked participants to write down their answers; instead, the organizers opened the door for free discussion. By asking participants to discuss the storage environment, the organizers wanted to approach the subject of whether or not it is worth waiting for Puer tea to age.

A caked Puer tea from 2001, known as China Tea Yellow Seal (Zhongcha Huang Yin; the character *zhong* was yellow on the package), was the first to be infused. Around the rectangular tea table on the ground floor, almost twenty people watched as the tea was infused. They observed its saffron color in the serving pitcher. Then tasting began. People silently sipped the tea from their tea bowls. No one talked. The participants' real identities were kept secret, since they all used their web names to sign in at the reception desk. Few introduced themselves or said much at the tea table. Perhaps the serious atmosphere was just what the tea appreciation called for, but it made the atmosphere cold and awkward. Mr. Wei, another key organizer of the event, tried to enliven the room by explaining what he knew about the general life history of the tea. But he stopped short, embarrassed by the lack of response. No one praised or criticized him, or stood up to disagree with his commentary, as people often did on the Sanzui website when the topic touched upon the mythical life history of Puer tea. Mr. Huang, who had been working hard at the computer desk, noticed the strange atmosphere when he returned to the tea table and called on people to speak up. Not receiving any response, he then urged Mr. Wei to continue with his commentary. Mr. Wei instead suggested that everyone should talk, but he was ignored.

On the first floor, Mrs. Fan, the owner of the teahouse, infused tea at one of the tea tables. While busying herself with the tea sets in front of her, she announced to guests that the tea was carefully selected from many contributions, and that the selections had been tasted by the organizers and deemed excellent. Moreover, she added, all of the chosen tea had been kept in "dry storage," proving that it had been stored in a clean environment.

(This was in comparison with "wet storage," in which the humidity could accelerate fermentation but also encourage the growth of bad bacteria, and hence was increasingly rejected by Puer tea collectors.) More information about the tea was disclosed, leaving room only for comments, agreement, or opposition from the guests.

Mr. Yan strengthened Mrs. Fan's point by moving between the tea tables on the two floors. He was a well-known judge of Puer tea and had tasted all six samples several days before. As the main organizer, he expressed his personal feelings about each tea, hoping to elicit more comments from the attendees. He asked people's opinions after his own statements at each table but received only brief comments, such as "Good," in return. At one point he tried to get opinions from several people who worked for a magazine. Journalists were usually considered good at asking questions and giving comments, but to Mr. Yan's disappointment, they didn't say much after he told them that the tea they were drinking came out of dry storage. Encountering reluctance and embarrassment, Mr. Yan's strategy was often to make an improvised joke, relevant to the tea or to the drinkers.

Throughout the tea-tasting event, which lasted about four hours, I noticed that most people remained silent. Voices often arose from the organizers or from people who were good at joking. Many exchanged private opinions in low tones with their acquaintances or those sitting near them. Few people publicly declared their opinions about the tea. This reticence took away from the tea event's significance for competition or discussion, and it formed a striking contrast with the disputes that occurred in the active battlefield of the website. Most participants were instead preoccupied with taking pictures of the tea, collecting and preparing something for the future rather than experiencing the present moment.

Based on my observations and conversations afterward with several participants, I concluded that the silence reflected people's concern with saving face (*mianzi*). As several people commented, tea can facilitate communication, but the pleasure of sharing tea is possible only when there are fewer participants. The Sanzui tea meeting, at which over fifty people were arranged in fixed seats, and which had a ritualistic atmosphere, contrasted with the old Chinese custom and was not conducive to communication. Many people felt too shy to express their personal feelings in front of such a large audience, and many were afraid to lose face confronting the sensitive and competitive issues around Puer tea. The experience of communicating on the website had taught people that they would be attacked fiercely if they dared to say something different from the general opinion or sneered at relentlessly for making a small mistake about certain facts.

Some were also concerned about saving the "face" of the organizers. When the second Puer tea was served, one participant bravely observed that the tea was a bit wet and might have traveled via air, implying that it must have been kept in wet storage (in Guangdong or Hong Kong) and flown from there to Kunming. But he didn't continue with his comments, perhaps because there were no other responses, and perhaps because he understood that his comments were contrary to the organizer's description of the tea.

These tea samples, though they were not actually produced by the organizers or contributors and had been passed from one collector to the other, had been endowed with something spiritual and become part of the organizer or contributor, like *hau*, the spiritual power of things in the Maori gift exchange (Mauss 1954). The tea symbolized the master's ability to identify the tea, his fortune in encountering it, his ability to possess it, and his courage to reveal it. In this sense, the place where the tea had been stored was identified with the social space of the person who contributed, selected, and spoke for it, so to criticize the tea was to criticize its master. In this way, commenting on the tea meant exposing one's "cultural capital" (Bourdieu 1989), which in this context became one's "tea capital," which might be in conflict with the "tea capital" of another participant. Criticism endangered the relationship between people (*guanxi*), especially in a face-to-face context. Improper words could make both the speaker and the master lose "face" (Kipnis 1995).

This concern with "face" also displayed an oppositional attitude. The famous post about *huyou* on the Sanzui website, which advocated remaining loyal to one's own innate senses rather than being swayed by external information, had been praised by the organizers. However, as the participants had witnessed, before and during the tea tasting, the organizers had given a great deal of guidance. That is, although the organizers didn't like being swayed by others, at their own tea-tasting events they were swaying their audience. In addition, although the organizers had originally set out the value of aging Puer tea as a subject for discussion and asked participants to freely discuss it, all the visual and oral information they provided actually indicated that the issue didn't need to be discussed further, for older Puer tea was surely better.

So, participants battled in ways other than speaking, like the heroes in martial arts novels contesting with inner force (*neigong*) rather than an outer force (*waigong*).[6] Their silence showed the attempts of the opponents to keep their distance from the organizers' interest. It also reflected their attitude toward the organizers and the difficulty they had in finding a proper social space at the event. Silence stood not only for people's distinct self-representations but, more importantly, for the way they related, interacted, and disagreed with one another.

THE SPACE OF BELONGING

As the Sanzui tea tasting continued, I gradually noticed that people became more likely to express themselves when a certain tea gave them a sense of belonging. This was most obvious when the tea tasting reached the fifth sample, Xiaguan Bowl Tea Grade B (Xiaguan Yi Tuo) of 1988.

Because I was focusing on filming the event, my sense of taste was lost, but when the participants commenced tasting the nineteen-year-old Xiaguan Bowl Tea Grade B, I observed through the camera something different in several people's expressions as they drank the tea. I was sitting at Mrs. Fan's table at the time. Mr. Zhu, the editor-in-chief of one of the popular Puer tea magazines, came over from the other tea table to chat with two friends and join their tasting. The group used gestures to enhance their appreciation of the tea, indulging in personal experiences and not paying attention to the directions they received. They sipped the tea with a look of pleasant surprise in their eyes; they smelled the empty tea bowl over and over; and they chewed the tea leaves, trying to experience the flavor more deeply. Then, for the first time at the tea meeting, I heard someone express his own taste impression directly. Mr. Zhu and his friends praised this tea highly and concluded that it was "vigorous" tea (*meng cha*), like the *kungfu* of Shaolin, which is played with sticks, in contrast to the seemingly softer Taiji style.[7]

According to the authoritative view of many tea experts, good Puer tea should be sun-dried; if it is machine or fire dried, it won't be suitable for long-term storage. Xiaguan Bowl Tea Grade B is well known for having been machine or fire dried by the Xiaguan tea factory; it was naturally fermented, though Mr. Zhu and his friends were unsure where. But they agreed that it had been kept in relatively dry storage, which they saw as a positive environment. Its excellent taste that day proved to Zhu that machine or fire drying isn't always bad and that good dry storage can play a significant role in directing the natural fermentation of Puer tea toward a pleasing result.

Before this event I had learned that Mr. Zhu was a raw tea connoisseur, someone who preferred to drink naturally fermented raw Puer tea, either young or old. He despised artificially fermented tea. Among the three kinds of raw Puer tea served at the Sanzui event, the 1988 Xiaguan Bowl Tea Grade B was the one he liked most. At that moment Mr. Zhu shared a similar interest with his two friends as well as with the organizers, who recommended and brewed this tea. By challenging the authoritative position, Mr. Zhu found a sense of belonging at the event, a pleasant social space in which he felt valued.

SEPARATE TEA-TASTING EVENTS

However, not everyone reacted like Mr. Zhu, reaching a moment of excitement and finding a feeling of personal belonging. On the contrary, quite a few people were displaced by the tea meeting, and all the teas served that day tasted bad to them. The case of Mr. Yang was typical. I met him at a tea-training class, where he taught children how to properly infuse tea using the *gongfu* technique as well as supervising them in traditional arts, such as Chinese poetry and calligraphy. The tea classroom was also his teahouse, where he drank tea with close friends. In the classroom space, he was well respected by the children, their parents, and his friends. He was praised as a highly professional tea master.

Mr. Yang showed up one hour late to the Sanzui event, where he signed in at the front desk and was directed to the first floor. Not more than an hour later he came downstairs and told me in a low tone, "I am leaving. It doesn't make sense to stay here. I'll hold a tea-tasting event at my tea classroom soon. Please come to enjoy it. It will be much better than this one."

About a week later, I joined in and heard more about Mr. Yang's complaints. He admitted that he should not have been late for the Sanzui event, but he thought the organizer should have assigned each participant a seat. He managed with difficulty to find a seat at the corner of one tea table, but no one noticed that he had just arrived and he was not promptly served a tea bowl. Even later, when he managed to get a bowl, he found it hard to reach the pitchers containing each brew and to communicate with others. Now, sitting in his own tea classroom, he acted as the infuser for ten people. He interpreted the Chinese tea law for me:

> What is interesting about tea is that it is a good medium for people's communication, as I am teaching children something via tea. If you have tea but lack good communication, the tea loses its significance, and I'd say that it has been infused unsuccessfully.

Mr. Yang served us more than six kinds of Puer tea. Unlike at the Sanzui event, in which electric kettles were used to boil the water, he used high-quality charcoal and a silver pot. In my opinion, Mr. Yang's tea didn't rival those served at the Sanzui event. But to Mr. Yang, they were clearly much more enjoyable. Where he had been unable to find a seat at the Sanzui event, he now sat in the master's seat. The tea classroom was precisely the social space where he had a sense of belonging.

Mr. Yang was not the only guest who felt uncomfortable at the Sanzui event. Mr. Ping was another. He went to the Sanzui event at the encouragement of his friend Mr. Zhu, the editor-in-chief of a Puer tea magazine. Unlike Mr. Yang, who had no acquaintances at the event, Mr. Ping sat with Mr. Zhu and some other friends from the media. He was greeted warmly by Mr. Yan, the organizer, because he was a well-known writer on Puer tea in Yunnan. He tasted all of the tea that day but left before the event formally ended. A month later, when I talked to him, he recalled the tea event as an unpleasant experience. He commented that it was just a kind of cultural pose (*fu yong feng ya*), and that all the tea there was worthless.

Having read Ping's works on Puer tea several years ago, I knew that he was also a raw Puer connoisseur, like his friend Mr. Zhu. But talking with him, I learned that he was an even stricter purist. He rejected not only artificially fermented Puer tea but also raw tea produced by large-scale tea factories. He was fond only of handcrafted raw tea from small-scale producers. Furthermore, he demanded to know where the tea had come from. For example, some of his Puer tea came from a close friend who was a tea master in Xishuangbanna. This friend went to the nearby tea mountains to pick tea leaves, which he processed himself. Ping regarded that kind of Puer tea as precious. By contrast, he viewed most of the Puer tea circulating in the market, and especially the tea that was provided at many conferences or ceremonies, as trash.

Ping also emphasized the contribution of the Yunnanese to Puer tea, in contrast to the dominant tea literature from central China and the dominant consumption trends of Puer tea in Guangdong, Hong Kong, or Taiwan. This made me realize that, to him, the pursuit for raw Puer tea was not only rooted in traditional Chinese ideas about nature but also linked to his personal identification as a Yunnanese. He regarded raw Puer tea as being more indigenous to Yunnan, whereas artificially fermented Puer tea was invented to respond to the consumption demands of drinkers in Guangdong and Hong Kong.

Even some who had not attended the Sanzui event made negative comments about the tea served there. Mr. Wen was an example. I met him when he came to Yiwu in the spring and autumn to trade tea. During the summer and winter, he sat in his own teahouse in Kunming, gathering with friends and sharing his achievements—the tea he traded and supervised for processing. He was a well-known member of Sanzui's online tea forum, but he didn't go to the event. When I asked him why, he said, "I think it's nonsense. I knew I would not like that tea even without going. I knew that some people would be there who are fond of directing others." Wen's teahouse was not open to the public, but was a private space for tea enthusiasts. Wen jokingly called his frequent guests *menke*—retainers who originated in China in the eighth century B.C.E., who were accommodated by the master and served him. Some *menke* came to Wen's teahouse every week or even every couple of days to drink tea without paying for it (fig. 7.2). But once Wen was back from the production area they bought tea from him to drink back home. He was respected by his *menke* as a tea master with knowledge that extended from field to table, and he was happy and able to answer many questions about tea.

By keeping silent or even not attending the Sanzui event, these critics displayed an interest distinct from the organizers of Sanzui. In order to declare their opposition, they held separate events.

Another case that is distinct from those discussed above involves Puzi (web name), one of the Sanzui organizers who attended the big event. Unlike Yan or the other organizers, he seldom spoke, though it was well known in the web circle that he had considerable tea knowledge. I met him in Xishuangbanna when he organized a small team to perform a field investigation of Puer tea. Back in Kunming I visited him many times. Unlike the organizers of the separate tea-tasting events discussed above that often

FIG. 7.2 Frequent visitors have their own tea bowls at the master's tea room. Photo by the author.

had many people and were held at independent teahouses, he preferred not to call his tea tasting a formal event. He sectioned off part of the living room in his house, simply but elegantly decorated, to serve as a tea room. Usually he invited only a few guests at a time, in accordance with Chinese tea custom: "Drinking tea alone makes a person focus on the spirit of the tea, but sharing with another is superior; three to four people together can share more pleasure in drinking tea, five or six is too many, and seven or eight is charity" (Chen Zugui and Zhu Zizhen 1981: 142). I knew that Puzi had certain principles: not to be directed by others, and also not to direct others. Puzi commented when he made tea in his own tea room that the supreme pursuit for Chinese to drink tea should be harmony (*he*), a concept in Chinese philosophy that includes tolerance of diversity or heterogeneity. The Sanzui tasting event looked peaceful because no debate happened there. But according to Puzi, it didn't really achieve harmony. His comments indicated that there actually had been many disputes about Puer tea that were played out secretly, fostering intolerance. For Puzi, the debates were often meaningless, and the most enjoyable way of tasting tea was just to share it with several good friends in his small tea room.

In martial arts novels, the Chinese knights-errant are generally those who live in the world of *jianghu,* subtly declaring their opposition to the court. Puzi's case made me think of some special knights who retreat from *jianghu* itself after learning enough and tiring of all the *jianghu* disputes. That is, not only do they go beyond the court but they also go beyond the *jianghu.* To them, neither the world ruled by the court nor the *jianghu* is a perfect society (Chen Pingyuan 1997: 176; Hamm 2005: 137–167). By keeping distance from both the court and the *jianghu,* they declare a singular interest different from that of all the other actors. But importantly, when they construct their unique social space, they also reconstruct the way that they relate to and interact with others.

CONCLUSION: MULTIPLE SPACES

The Sanzui tea-tasting event ended after about four hours. The guests left and the organizers stayed for an informal summary. Mrs. Fan opened the discussion by scolding Mr. Yan for failing to organize the event well. In her opinion, Mr. Yan should have asked each participant to give an introduction at the beginning. This would have facilitated communication during the tasting. However, she said, Mr. Yan enjoyed chatting only with particular people, and, as a result, most guests remained unknown to one another and to the organizers. Few comments on the tea were collected. Mr. Yan didn't deny this. He was tired, and he hoped that there would be feedback on the website to make up for the lack of comments in the actual tea-drinking space.

As anticipated by Mr. Yan, the discussion of taste and feelings about the aging of Puer tea took place mostly in cyberspace. The night of the tea tasting, he wrote a new post, attaching his own comments and reminding participants to add their own hundred-word comments. Some people responded at once. Several wrote comments very carefully, tea type by tea type, drawing on their memory and illustrating their comments with photos. Behind their anonymous web names, they began to comment more bravely. Generally, they confirmed the value of the six types of tea, but some people doubted the age of certain teas and voiced suspicion that some had been put in undesirable "wet storage." These alleged shortcomings generated immediate debate. One person didn't give detailed comments but just said that most of the teas were terrible. This comment met with a strong rebuttal, and its author was asked to set out his criteria for good tea. The website became a battlefield once again, in contrast to the seemingly peaceful world of the teahouse.

Multiple layers of space were touched upon through the Sanzui event. First, there was the tea storage space that was the topic of discussion for the event. Second, there was the teahouse where the tea tasting was held. It should have been here that participants had a full discussion about storage space, although in fact that discussion went on mostly silently and unsuccessfully. Third, there was cyberspace, originally envisioned only as a supplement to the real teahouse for collecting comments. In fact, this became the space where more "real" debate about tea storage went on, illustrating how the Internet is being used in China as an important site for debate. Fourth, there were other teahouses or tea rooms where separate tea-tasting events were held. Although representing a stance of separation or retreat rather than direct contest, they showed a much stronger counterattitude than did the virtual space toward the large event. Finally, there were the symbolic social spaces of different groups and individuals. These social spaces were distinct but also interacted.

Participants in the tea-tasting event should have paid attention to the question of tea storage space, but they in fact neglected it almost entirely; they instead cared more about human interaction. As the attention of these participants shifted from tea space to human space, their tea-tasting skill became less important than their interpersonal communication.

CHAPTER 8

Interactive Authenticities

The enchantment of Puer tea lies in its unsolved mystery, its vagueness, and its changeable meanings. It's like a sea, vast and mighty, filled with submerged reefs and strong rapids, and no one can reach its far end. What we can do best is just sip our own tea!

—Yang Kai, Liu Yan, and Li Xiaomei 2008: preface

Zongming, a friend from Hong Kong, visited Kunming in December 2007, seven months after the price of tea plummeted. Making use of his holiday, he met in Kunming with friends he had communicated with on the tea website Sanzui. The participants on Sanzui lived in almost every province of mainland China, especially in urban areas. After the Puer tea recession, many participants traveled between Yunnan, the production region, and the Pearl River Delta, the so-called consumption area, to visit one another. Each side was eager to know what was happening in the other location, and both wondered about the future prospects of the Puer tea market. The Sanzui participants in Kunming gave Zongming a warm reception, as they were excited to meet someone who was known as a good commentator and who might bring some new information from Hong Kong. Because Zongming was not a trader but just someone with a keen interest in tea, his Kunming tea friends, many of whom were tea traders, talked openly with him without worrying too much about commercial competition.

I got to know Zongming in Hong Kong one year before his visit to Kunming, when he took me to several *yum cha* restaurants to show me how Hong Kong people drank Puer tea in their daily lives. In Kunming, I joined him on some of his trips to teahouses. When issues about consumption are raised that refer not only to Puer tea's production but also its storage, the social biography of Puer tea becomes more complicated and contested. The various features of aged Puer tea reflect the changeable social landscapes in consumption, exemplified by the temporal contrast before and after China's Reform and the spatial differentiation between Yunnan (the production area) and Guangdong, Hong Kong, and Taiwan (the consumption areas).

Understanding a commodity's circulation may be approached by examining its detailed social biography in exchange, rather than focusing only on its exchange forms. Thus what links a commodity's value and exchange is politics—namely, "the constant tension between the existing frameworks (of price, bargaining, and so forth) and the tendency of commodities to breach these frameworks" (Appadurai 1986: 57). The value of a certain commodity, its path of circulation, the knowledge it contains, and the desire and demand for it are all determined by social definitions and redefinitions, and hence the authenticity of things cannot be static but shifts contextually. The tension around Puer tea, too, may be seen as the contest between multiple self-presentations across time and space. On the one hand, the "habitus" (Bourdieu 1984, 1989) of consuming a certain type of Puer tea is shaped by a certain nature and culture, which shows a strong identification with localization. On the other hand there is a global intent to control, to provoke, and to import capital to the local. In order to cater to the outside demand, the localized "habitus" is forced to adjust and reach a certain compromise with globalization. While compromising, the local forces are also redomesticating the outside forces to serve in the local's new self-presentation. Thus it becomes neolocalization, in which the global and the local elements are mixed and become hard to differentiate, and one's self-presentation is never self-determined, but actually involves the borrowed, adapted, and reauthenticated elements from others, as has been mentioned by previous studies on consumption.[1] But such neolocalization is never finished, because there are always new forces, whether from the global or the local, to further challenge the existing authenticity of Puer tea, like the *jianghu* battle in which new risks and disciplines always emerge to break the old format. So, situated in the transformation of China, when old concepts meet new desires, and located in a *jianghu* contest, Puer tea acquires multiple versions of authenticity.

This "multiple" perspective not only recognizes the power of localization to cope with globalization but also displays an interplay between localization and globalization that often goes on with endless counterforces, shaping a changeable and varying authenticity for things as well as for people's social lives.

COLD AND WARM

The first friend Zongming met in Kunming was Hongtu (web name), who Zongming understood through web communications to be a defender of Yunnanese culture. For instance, on the Sanzui website there was a post by a participant from Guangdong titled "Puer tea doesn't need the Yunnanese." This post said that Yunnan was simply the area producing the basic material of Puer tea but that the people of Yunnan didn't contribute to Puer tea's trade and consumption as much as the Cantonese. The post was fiercely attacked by Hongtu, who enumerated many facts about the Yunnanese contribution to tea. To him, these great contributions had long been masked because Yunnan was remote from the political and economic center of China. Hongtu's strong identification as a Yunnanese could also be seen from his full web name, Hongtu Lantian, which means "red earth and blue sky," a popular description of Yunnan's natural landscape.

To entertain his honored guest, Hongtu brewed his favorite tea, a ten-year-old raw Puer tea originating in the tea mountain of Mengku. Mengku is located in Lincang, a southwestern subdistrict of Yunnan bordering Burma (see map I.1).[2] In recent years it had emerged along with Yiwu, Menghai, and several other places as a famous production site for Puer tea. According to Hongtu, Yiwu tea tasted too weak and Menghai tea was barely acceptable; only Mengku tea was enjoyable, full-bodied, and lingered long enough on the palate.

This particular tea had been aged in Kunming. In addition to Hongtu, Zongming, and myself, there were three other guests, all frequent visitors to Hongtu's tea shop and supporters of Mengku tea. Having gotten used to the taste of Yiwu tea during my fieldwork, I found the Mengku tea scarcely palatable, with a less subtle combination of sweetness and bitterness than Yiwu tea. I found that the Mengku fans at this tasting used the same language of praise as Yiwu supporters did: "The tea of Mengku/Yiwu is the remarkable flag of Puer tea," or "If you want to learn about Puer tea, you must first practice drinking and understanding the authentic tea of Mengku/Yiwu."

Zongming also declared his fondness for Mengku tea, but he didn't give it the same praise as the others. To him, the more problematic issue at that moment lay in the difference in flavor of teas not between different production areas, but between different storage places. The Mengku tea brewed by Hongtu was said to have been stored in Kunming for ten years, but in

Zongming's opinion it had not been sufficiently aged. He thought it was still too raw and far from smooth. To tea drinkers from the Pearl River Delta, smoothness was an important property, and they thought it resulted from storage in a relatively humid place, such as Hong Kong or Guangdong. For them, good Puer tea needed to be smooth in the throat when it was swallowed—as smooth as the slowly stewed soup (*lou fo tong* for Cantonese pronunciation; *lao huo tang* for standard Chinese Pinyin) commonly eaten as part of their daily meal. Drawing on ideas from traditional Chinese medicine, they argued that smooth Puer tea was warm for the body. By contrast, raw Puer tea was too irritating; it was intrinsically cold, and hence harmful to one's health (see Anderson 1980).

Zongming's response to this Mengku tea reminded me of a scene I had witnessed in Yiwu. In April 2007, I met a group of travelers from Guangdong who were visiting a Yiwu family who produced Puer tea. The family master brewed some recently made raw tea, a superior type according to him, to entertain his guests. The guests, however, felt nervous about this fresh tea. They sipped only a tiny bit from each run. At the third run they asked the master to stop and suggested that he brew the aged tea they had brought from Guangdong. One guest told me that he felt his heart pounding when he tasted the raw tea. Nevertheless, in the end, all the travelers bought a large quantity of raw Puer tea from the local family. Perhaps the "adventurous" raw tasting had made them foresee a good prospect for the fresh tea, hopefully via storage back in Guangdong.

Zongming, who had tasted various kinds of Puer tea, was not nervous in the face of the Mengku raw Puer tea. But like the Guangdong travelers in Yiwu, after tasting the Mengku brew, Zongming asked if he could infuse a Puer tea he had brought from Hong Kong, in order to show his preference. It was a twenty-five-year-old tea packed in a bamboo pipe that had been stored in Hong Kong. Its brew was darker than that of the Mengku tea. According to Zongming, it had reached a good degree of smoothness, had the medicinal smell that results from good Hong Kong storage, and was warm and beneficial to one's health. Now Hongtu found it hard to comment. After a long silence he said that the Hong Kong tea's smell was indeed special, but it faded once the tea was swallowed and couldn't be recalled until the next sip. He also remarked that the tea didn't have a long aftertaste, a property of that was very important to him. The other guests also commented on this tea's "strange" taste. They were trying to appreciate this twenty-five-year-old tea, and although they did not dislike it, they obviously didn't think it rivaled the Mengku tea.

In my experience, most Yunnanese, especially frequent tea drinkers, prefer raw and naturally fermented Puer tea, and they often have a preference for tea produced on a particular mountain—for instance, Yiwu or Mengku. Furthermore, they prefer tea that has been aged in Yunnan rather than elsewhere. Like Hongtu, they appreciate the lingering aftertaste of raw Puer tea. People from the Pearl River Delta, however, prefer Puer tea that has been stored in Guangdong or Hong Kong for at least five years. This aging, they think, creates warmth in the stomach as well as smoothness in the mouth. Such "standard taste" or "collective taste preferences" are shared by groups of people living in the same natural and cultural environment (Ozeki 2008: 144–145) and become the standard against which people judge other tastes.

Popular writers have increasingly argued that Puer tea could not be properly fermented until it was exposed to sufficient humidity and heat (Bu Jing An 2007). Hong Kong and Guangdong are close to the sea and are more humid than Kunming, which is located on a plateau.[3] Accordingly, some people, mainly Cantonese, argued that Puer tea should be stored in the Pearl River Delta after production in Yunnan. Some even said that five years of storage in Guangzhou or Hong Kong was equivalent to more than ten years of storage in Kunming. They believed that Yunnan had excellent tea resources but was unsuitable for storage or that the Yunnanese hadn't known enough about storage, even though Yunnan also had humid areas, such as Jinghong in Xishuangbanna.

After staying in Kunming for only one week, Zongming became sick, despite the warm and sunny weather. He began to cough and even vomited one day after eating spicy Dai food with Hongtu and several other friends. At that meal, he witnessed the Yunnanese capacity for eating spicy food. It seemed that the more pungent the food was, the more Hongtu enjoyed it, although he was sweating and his face was turning red. Another Yunnanese participant, Puzi, ate chiles quietly without his face changing color at all. At the start of the meal, Zongming tasted everything out of politeness, but soon he selected only the less spicy dishes. He didn't eat much, but he still suffered from the pungent food and had to use a lot of tissues.

A few days later, in Hongtu's tea shop, Zongming attributed his sickness to "water and earth not fitting" (*shuitu bu fu*). The dry climate of Kunming compared with the humidity of Hong Kong was one factor. Zongming also confessed that he persisted in the Hong Kong habit of taking a cold shower every evening.[4] This was contrary to local custom and strongly criticized by his Kunming friends.

Puzi made this point in another way. He had recently stayed in Guangdong for several months and said he could not bear the Cantonese food at first. It was too oily, and it had no flavor. He could not get used to the Puer tea stored in Guangdong, either. He described it as "like having Chinese medicine rather than tea." However, he soon found that he wanted more of this kind of Puer tea after a meal, as its medicinal flavor did, in fact, help him digest the oily food. In turn, after his digestion improved, he ate more, which then caused him to drink even more Puer tea. In the end, Puzi realized that he had grown to like Cantonese food and also the Puer tea stored in Guangdong, which he found to be complementary.

Zongming nodded his head as he listened, hearing in Puzi's story the counterpart to his own. Both of them agreed that the only way to get used to the local environment was to eat the local food and drink the local drink. Considering each other's position, they found that neither Puer tea was uniquely authentic, and they realized that one's preference between raw/cold Puer tea and aged/warm Puer tea was a matter of local culture. As Zongming moved from Hong Kong to Yunnan, he discovered that the authenticity of Puer tea was mobile, too. But he later found out that the authenticity of Puer tea not only diverged from place to place but also varied in the same place over different periods. Once it was tested on the historical timeline, the authenticity of Puer tea would become even more mobile.

PUER GREEN TEA: *FAN* AND *MI*

Hongtu's tea shop became a place where Zongming met more friends and tried more types of tea. One afternoon he met Lao Li, who was a tea trader in Kunming. Lao Li was about the same age as Zongming, and the two shared many comparable experiences. On this afternoon, Hongtu took out a special tea to brew. It was special not because of its age (fifteen years), nor because of where it had been stored (Yunnan), but because of the name marked on its box: "Puer green tea made from spring buds" (*Puer lü cha chun rui*). It was loose tea, and the term "spring buds" proved that it was made of the highest-grade raw material. What was confusing was that it claimed two identities: Puer tea and green tea (fig. 8.1). Puer tea had

FIG. 8.1 The packing of Puer green tea. Photo by the author.

originally been categorized as dark tea by academic tea experts in China. Around the turn of the twenty-first century, some tea researchers from Yunnan began to argue that it should be classified in an independent tea category because its production process was distinct from that of dark tea. Before the boundary between dark tea and Puer tea could be established, a new dispute over the distinction between green tea and Puer tea emerged. Loose-leaf tea made from spring buds had long been considered a variety of high-quality green tea in Yunnan, but by 2007, when Hongtu infused it for Zongming and Lao Li, it had come to be recognized by the market as a kind of Puer tea despite its misleading name. The reclassification happened after Puer tea became famous, and this newfound recognition would increase Puer green tea's value. If it was green tea, after fifteen years it would have lost its value and could just be thrown away. But if it was Puer tea, it could be stored for a long time and sold for a high price.

Whatever it was, Hongtu infused it for the guests. The brew was reddish-yellow, proving that some transformation had occurred during aging. Lao Li, who had tasted a similar kind of tea when it was fresh, pointed out that the brew would be faint yellow if it hadn't been aged. I found the flavor astringent, not as smooth as the bamboo tea, without the lingering aftertaste as the Mengku tea, and lacking the fresh aroma that one would expect from green tea. Lao Li said he agreed with Zou Jiaju, the head of the Yunnan Tea Association, who advocated in his books and on his blog that tender tea buds were good for making green tea but not good for Puer tea; tough tea leaves and stems were better for Puer tea and aging would only improve them further (Zou Jiaju 2004: 90; 2005: 133). Zou's idea had many opponents, and another of his ideas encountered even more opposition. He argued that newly produced raw Puer tea that had not been further fermented was not really Puer tea, but was in fact closer to green tea. He referred to such tea as *mi* (raw rice), and authentic Puer tea as *fan* (cooked rice). The latter category included both aged Puer tea with long natural fermentation and Puer tea with accelerated artificial fermentation (Zou Jiaju 2004: 4; 2005: 109).

Zou was from Yunnan. Working in the Yunnan Tea Import and Export Company had given him many opportunities to learn about the tea-drinking habits of the Pearl River Delta, and in Yunnan he famously promoted artificial fermentation. When I interviewed him, I found that he had accepted the idea of *fan*. While we were talking, he brewed the artificially fermented Puer tea that was produced by his company. Later we went out for lunch. The restaurant we went to is known for its thin rice noodles, but Zou chose the thick ones. When I asked why, his answer was consistent with his Puer tea preference: "The thick ones are fermented, but the thin ones aren't."

Raw unfermented Puer tea is the dominant type consumed in Yunnan, but some Yunnanese prefer the fermented type. Some tea drinkers attributed this preference to the condition of their stomach. They agreed with Zou Jiaju, and also with people from the Pearl River Delta like Zongming, who felt that *fan* was warm, smooth, and beneficial for one's stomach, but *mi* was too astringent and harmful, and should be consumed only in small quantities. Furthermore, I found that people who had been trained in tea science, like Zou Jiaju, stressed that the key characteristic of Puer tea was "postfermentation." That is, there had to be a certain degree of chemical reaction in the tea, whether oxidation or the biochemical reaction caused by microbes. As far as they were concerned, Puer tea could be naturally fermented, but this took a long time, whereas artificial fermentation took only two or three months. Because aged raw Puer tea was not available in Yunnan—most of it having been stored in Taiwan or Hong Kong—the artificially fermented type became the reference point for the tea science people of Yunnan, who regarded artificial fermentation as a great innovation in tea processing (Liu Qinjin 2005; Xu Yahe 2006; Zhou Hongjie 2004; Zou Jiaju 2004; Ruan Dianrong 2005a).

However, this viewpoint was opposed by some producers who didn't take up the artificial fermentation technique, and by some traders whose business concentrated on newly produced raw Puer tea. It was also opposed by some consumers, mostly from Yunnan, whose palates were faithful to the raw type, which they didn't consider harmful to their health. In other publications and websites, and in informal conversation, these opponents countered that Zou was abridging the history of Puer tea. They said that according to Zou's authorization of *fan,* the history of authentic Puer tea could start only in 1973. They questioned how to define the compressed tea freshly consumed in Yunnan a long time ago, which was carried to Tibet and to the emperor in Beijing during the eighteenth and nineteenth centuries (Gao Fachang 2009: 193–194).

The ambiguous "Puer green tea" served by Hongtu and the debates between *mi* and *fan* showed the complexity in distinguishing the authenticity of Puer tea in recent years. The preference for *mi* or *fan* was not simply based upon acclimatizing to local climate and customs, as discovered by Zongming. Preferences were also shaped by the negotiation of trade and consumption between Yunnan and the outside world. The preference for warm and smooth Puer tea from the Pearl River Delta was an outside forces, which had influenced the categorizing of Puer tea in Yunnan and also inspired the innovation in production techniques. At the same time, there were internal forces, such as the accelerating economic development of Yunnan and its desire for distinct self-representation. This self-representation, shaped by negotiation with the outside, presented multiple rather than singular voices.

When I asked how long they'd been drinking Puer tea, most of my interviewees from Yunnan told me that they had been doing so since the mid- or late 1990s, or sometime after 2000. No one gave a year prior to the mid-1990s, suggesting that a popular and "clear" conception of Puer tea didn't emerge in Yunnan until after that time and especially after the turn of the twenty-first century. When I asked the same question in Hong Kong, most of my interviewees found it hard to answer. Many of them said they had always drunk Puer tea. One elderly man whom I met at a *yum cha* restaurant told me, "I know that before I was born, my grandfather drank Puer tea."

Back at Hongtu's teahouse, Zongming and Lao Li, both fifty-three, declared that they had grown up drinking Puer tea, one in Hong Kong and the other in Yunnan. The discussions about "Puer green tea," however, made it clear that the two men had grown up with different versions of Puer tea. To Zongming, the Puer tea drunk daily with *yum cha* in Hong Kong was *fan,* the artificially fermented or aged raw type. The transformation of Puer tea from *mi* to *fan* is said to have been casually and "naturally" accomplished in the long journey by caravan from Yunnan, or via aging in storage by the Cantonese. These two factors seemed more plausible when they were combined with another story: from the 1950s to the early 1970s, unaged Puer tea that was exported by rail and road from Yunnan to Hong Kong was rejected by local consumers, who were reluctant to recognize such raw tea as Puer tea (Zhou Hongjie 2004; Zou Jiaju 2004). Rather, they preferred to wait until this raw tea was fermented in Guangdong or Hong Kong with added water (*fa shui*) and blended with tea leaves from Guangdong. The rejection of the raw tea was due to the fact that modern transportation had replaced the slow caravan. Many months' "cooking" by caravan was suddenly shortened to only two or three days. Under pressure from these changed conditions, the national tea factory of Yunnan sent its staff to learn the basic fermentation techniques from Guangdong and Hong Kong, and finally a more mature technique of artificial fermentation was invented in Kunming in 1973 (YTIEC 1993; Zhou Hongjie 2004; Zou Jiaju 2004). With this new technique, tea was transformed from *mi* to *fan* in just two or three months.[5]

The situation within Yunnan, however, was more complicated and changeable. As the production region, it had to cater to the demands of outside consumption, but at the same time locals swung between preferring *fan* or *mi*. While producing *fan* for Hong Kong consumers, the Yunnanese also continued to produce *mi* to consume themselves. To Lao Li, the teas that he had grown up with before 1973 were a variety of locally produced green tea. The difference between these teas depended mainly on which drying process had been used: sunning, steaming, stir-roasting, or baking. *Mi*, the uncooked Puer tea in Zou Jiaju's terms, actually referred to Yunnan sun-dried green tea, which was also at that time called Dian Qing.[6] Lao Li told us that the concept of Puer tea then was very vague. Normally the compressed form, such as bowl-shaped tea, was called Puer tea (actually raw tea), while the loose-leaf tea was sometimes called Puer tea and sometimes called green tea. The perplexing name "Puer green tea," Lao Li said, could have come from this period of vagueness. No matter what, though, if the green tea was over two years old, it would be thrown away. This Puer green tea, according to Lao Li's supposition, survived because of the carelessness of someone who had forgotten it, as storing Puer tea for its aged value wasn't practiced in Yunnan until recently.

Lao Li said that even after artificial fermentation was invented in 1973, he didn't immediately have a chance to taste the new artificially fermented Puer tea. This further convinced me that this artificially type of Puer tea was not made for domestic consumption in Yunnan. Rather, it was intended "to cater to the demand of the international market"—namely, Hong Kong and Macao, where Puer tea was consumed and also traded to other Southeast Asian and European countries (YTIEC 1993: 160). Lao Li first encountered artificially fermented Puer tea in the mid- or late 1990s. At first, its "moldy" smell surprised him and he wondered what type of tea it was. But he gradually came to accept its flavor as well as its function of warming the stomach. He also accepted that it was Puer tea, as he had learned from the market and some popular books.

Why did Yunnanese people start drinking artificially fermented Puer tea in the mid- to late 1990s? This can be explained in terms of China's economic development. The mid-1990s marked the beginning of China's economic ascent after the Reform and Opening Up. All kinds of material culture from overseas, such as pop music and food from Hong Kong and Taiwan, were entering the mainland and becoming popular (see Gold 1993). Yunnan, as a relatively undeveloped southwestern province, was urged to accelerate its economic development under a major national project called "Opening the West" (*Xibu da kaifa*), formally proposed by the central government in the late 1990s. It was in this context that Yunnan tried to boost its economy. Tourism to Yunnan began to increase at that time, and souvenirs were needed. Puer tea became one of them. One tour guide in Kunming told me that in 1996 and 1997, when she took tourists to tea shops in Kunming, what passed for Puer tea—a local and uniquely Yunnanese product—was always dark-colored, artificially fermented tea. So, artificially fermented Puer tea consumed in Hong Kong and Taiwan (which bought Puer teas from Hong Kong) began to enter the view of consumers in Yunnan, who suddenly realized that this popular "overseas" drink was originally locally produced. And in fact, at this stage, "Puer tea" referred to *fan* (artificially fermented Puer and aged raw Puer), whereas *mi*, then called Dian Qing, was applied to green tea.

Lao Li recalled that in the early 2000s, four or five years after he started drinking artificially fermented Puer tea, the market began to differentiate between raw and artificially fermented Puer tea. Dian Qing (Yunnan sun-dried green tea) was now considered Puer tea, but was given a new name—raw Puer tea. Tea advertising encouraged the aging, rather than fresh consumption, of both raw and artificially fermented Puer tea. The inclusion of raw Puer tea resulted from cooperative action between the local state and traders to promote Puer tea output in Yunnan (Tang Jianguang, Huan Li, and Wang Xun 2007a). Promoters of Puer tea recognized that artificial fermentation required special techniques and had high investment costs whereas raw Puer tea was easier to produce in large quantities and "natural fermentation" could be left to consumers themselves. That is, *fan* was hard to cook, whereas *mi* was easy to supply. This approach happened to meet the original palate of some Yunnan consumers for green tea, with which raw Puer tea had a close relationship. At the same time, the Taiwanese demand for raw Puer tea also helped boost its popularity. I learned from interviews with Hong Kong and Taiwanese traders that, as in Yunnan, many Taiwanese initially disliked the "moldy" smell of artificially fermented Puer tea and preferred naturally fermented raw tea. As a result of these various influences, in the first years of the twenty-first century, Yunnan began to promote the production of raw Puer tea. In this period many popular books, by both Taiwanese and Yunnanese writers, favored raw Puer tea over artificially fermented Puer tea, as it was aged naturally and was thus considered healthier and richer in cultural meaning (Lei Pingyang 2000; Deng Shihai 2004). It was even rumored that artificial fermentation could cause cancer. Yet the debate between *fan* and *mi* continued, and some tea experts, such as Zou Jiaju and some researchers from the tea academia, acknowledged only *fan* as true Puer tea. They stressed that the central feature of Puer tea was "postfermentation," and they argued that the microbe generated during artificial fermentation was beneficial to people's health (Liu Qinjin 2005; Zhou Hongjie 2004, 2007; Zou Jiaju 2004, 2005). Nevertheless, there was general agreement that both *mi* and *fan* could be aged for a long period. The old habit of throwing away tea after two years was largely abandoned. "The older the better" became the basis for the ongoing popularity of all Puer tea.

NEW TRADITION

When we tired of sitting around the tea table, Hongtu, Lao Li, and I took Zongming for a walk around Kunming. This trip allowed Zongming to see how the updating of what was considered authentic Puer tea had occurred alongside urban development in the city.

We started at Hongtu's tea shop. After walking for only five minutes,

Zongming noticed that there were many tea shops nearby. Hongtu's tea shop is located beside Green Lake Park, a central leisure place in Kunming. Surrounding the lake, which is about 2.5 kilometers long, are almost thirty establishments that serve tea. Five more are located inside Green Lake Park. Hongtu's tea shop is an example of one type of tea establishment, which sells tea—mainly Puer tea—along with tea sets and associated decorations. Guests can sit down, talk to the tea master, and participate in a free tea tasting before actually buying. But seats are limited and available for only seven or eight people at a time. Another type of tea establishment is larger and has ample seating. The main service in this kind of shop is to provide infused tea for guests to enjoy by themselves. Some also supply snacks, juice, or even wine. This is the most common type of tea shop around the lake. The least common, and the most remarkable in appearance, is the third type, which serves as both a restaurant and teahouse, housed in quadrangle courtyard dwellings, mostly built in the late nineteenth or the early twentieth century by wealthy people. With tiled roofs, wooden pillars, wooden floors inside, and stone paving in the courtyard, these shops share many similarities with traditional houses in Yiwu, but having been newly decorated, they look more luxurious. Formal lunch and dinner are provided, accompanied by tea or alcoholic drinks. Sometimes there are zither music performances. Between meals, guests can also come just for tea.

Kunming is the most important center for Puer tea distribution and consumption in Yunnan. The area around Green Lake Park is only one of the famous places where retail tea shops and tea restaurants are located. For wholesale tea trading, there are nine big wholesale tea markets, scattered around the urban fringe. The earliest one opened in 2002, and the largest one contains almost six hundred tea shops. At the end of 2007 four or five more wholesale tea markets were rumored to be under construction, but there were doubts about whether they would be able to successfully open due to the sudden recession in the Puer tea market.[7] According to one survey, at the end of 2006 there were a total of four thousand wholesale, retail, and service tea units in Kunming.[8]

Zongming was amazed at the number of tea establishments in Kunming. Though Puer tea consumption had long been more important in Hong Kong, there are no tea markets in Hong Kong that come close to the scale of the markets in Kunming. Most retail or wholesale specialized tea shops in Hong Kong are concentrated in a single commercial district, and there are only around ten shops total. Most Puer tea is consumed cheaply and in large quantities in Hong Kong's numerous *yum cha* restaurants.[9] When Zongming commented on this contrast, I recalled the words of an old tea trader in Hong Kong, who said as he handed me a cup of Puer tea brewed from a big porcelain pot in his office, "Sorry, we Hong Kong people don't drink Puer tea as exquisitely as you Yunnanese." In other words, he was not using the sort of delicate tea set and sophisticated method of tea infusing that was now popular in both the cities and tea production areas of Yunnan. I was amused by this comment, and I politely replied that Yunnan had learned a lot from Hong Kong about how to drink and store Puer tea. The Pearl River Delta, especially Hong Kong, was a developed economic region, whereas Yunnan had long been known as a lagging and poor area. But whereas Puer tea was routinely and "quietly" consumed in Hong Kong, it was treated in a more sophisticated manner in Yunnan. Zongming's surprise at the number of luxurious teahouses in Kunming echoed the old man's comment, but Hongtu, Lao Li, and I understood that this treatment of tea had become popular in Kunming only since the turn of the twenty-first century.

A social survey undertaken during the period of the Republic of China (1912–1949) recorded the presence of 350 teahouses in Kunming (Chen Zhenqiong 2004).[10] Teahouses were categorized into four types at that time. "Teahouses for pure tea drinking" (*qing yin chaguan*), which served nothing but tea but sometimes allowed peddlers to sell snacks, made up 90 percent of the teahouses. "Broadcasting teahouses" (*boyin chaguan*) provided music along with tea. "Pure singing teahouses" (*qing chang chaguan*) featured musical performances in addition to tea. And at "telling stories teahouses" (*shuo shu chaguan*), tea drinking was subordinated to storytelling. This last type was the one that people in Kunming were now most interested in recalling, as it combined two forms of typical and traditional Chinese entertainment. All four teahouse types functioned as places not only to quench thirst but also to socialize, or for individuals to enjoy a bit of "quiet" time amid noisiness, perhaps all day. The survey also showed that the dominant tea consumed in these teahouses was green tea produced in Yunnan. Puer tea was mentioned, but its definition at the time was vague, referring roughly to good-quality green tea.

The years from the 1950s to 1970s saw a decline in the number of these public tea places, both because of the nationalization of privately owned business after the establishment of the People's Republic of China in 1949, and because of the economic difficulties China faced, especially during the early 1960s. During the Cultural Revolution (1966–1976), political struggle became the dominant political theme and anything related to consumption was condemned. Lao Li was born in 1955. He recalled accompanying his father to storytelling teahouses when he was less than ten years old. He clearly remembered that most teahouses like that were closed when the Cultural Revolution began and when "Destruction of the Four Olds" (*po si jiu*)—meaning old ideas, old culture, old customs, and old habits—was advocated. Teahouses that featured storytelling, singing, or broadcasting were certainly considered old and undesirable, and Lao Li told us that drinking in teahouses at that time was despised as a luxury activity. As a result, tea was imbibed mostly at home or in the office and bought from a few national or collective grocery stores. The dominant type of tea at this time was still the green tea of Yunnan, which could be sun-dried, stir-roasted, or baked.

After the Reform and Opening Up in the late 1970s and early 1980s, traditional public teahouses together with other forms of entertainment, such as storytelling, were gradually revived in Kunming. Lao Li recalled that one shop that specialized in selling tea was opened in the commercial center of Kunming. This was the period that I could start recalling along with Lao Li. As a local Yunnanese born in the mid-1970s, I remembered that both at home and in public, the green tea of Yunnan was the dominant type, brewed simply in a glass. But Puer tea was known to be an indigenous product, and would be given as a gift to people from other provinces. However, its image was still vague at this time, and the term was mostly used to refer to compressed types of tea.

During the mid-1990s, public teahouses with "modern" style sprang up in the city center, such as around Green Lake Park. By modern, I mean that Western-style elements, such as tables, sofas, and curtains, were used in decorating them. These teahouses usually played popular music, both Western and Chinese, and served ice cream or juice in addition to tea. Some of these teahouses, like the major ones around Green Lake, were more like bars or cafés, but they were still called teahouses. They were often located in newer buildings with higher prices, and they were more commonly patronized by younger people, while the traditional teahouses were located on older streets and usually drew older customers.

Since the late 1990s, there have been bigger changes as well. An international horticultural exhibition was held in Kunming in 1999. Around that time, the city government launched a master plan to build a "New Kunming." Old streets and old houses were torn down, which in turn led to a reduction in the number of traditional teahouses. However, at that moment of destruction a new style of "traditional" teahouse and tea restaurant (like the third type of teahouse around Green Lake) emerged in Kunming. At the same time, Puer tea, an old local product of Yunnan that had been endowed with new meanings, became the dominant drink in all of the tea shops in Kunming.

Often, something that seems old is the result of recent invention, and "invented tradition" may be a part of the discourse of nationalism (Hobsbawm 1983). The recent interest in Scottish kilts, tartan patterns, and bagpipes, for example, shows that such nationalist passion inevitably follows a process of commoditization (Trevor-Roper 1983). Likewise, the emergence of Puer tea's popularity is the result of tacit and selective repetition with the past, an example of "adaptation [taking] place for old uses in new conditions and by using old models for new purposes" (Hobsbawm 1983: 5). It is through such reinvention that Puer tea is now proclaimed to be the very tradition of Yunnan, even though its authenticity has been continually up for grabs. Moreover, the new meanings about its tradition are strategically applied to convince consumers of a "worthwhile" investment. In this regard, Puer tea is discursively constructed and symbolized toward a localized nationalism, which intrinsically has the same flavor as that of commoditization.

Studies discussing the same period of spatial change in Kunming prove that the power of the post-Mao state was not weakened at all but rather reinforced through spatial reconstruction (Zhang Li 2006). The diminishing of old streets and the restructuring of Kunming since the late 1990s was implemented by the local authorities and the real estate sector in the name of "progress." That is, they wanted to make Kunming a truly modern city in an undeveloped region of southwest China, to escape the reputation of being behind and to catch up with national and international metropolitan areas. The substantial purpose behind such reconstruction was to "accumulate political capital and/or financial profits" (Zhang Li 2006: 464). We can infer from this that spatial forms are never natural but contain the power of "social control and political ordering" (Zhang Li 2006: 462).

Puer tea became prevalent against the same backdrop. The traditional teahouses that emerged were new in terms of when they were established but "old" because they borrowed numerous traditional elements. Some of them were established to "protect tradition," being located in old but redecorated houses. Ironically, this protection happened after the destruction of most of the old streets, and meanwhile the destruction and reconstruction transformed not only the spatial outlook of Kunming but also the "very modes of social life, local politics, and cultural identities" (Zhang Li 2006: 461). In the reconstructed traditional teahouses, people drank Puer tea, which had recently been recategorized according to the updated definition of Puer tea in the market. Since its resurgence, Puer tea had been promoted as "a drinkable antique," something that evoked modern nostalgia. The new "traditional" teahouses provided a niche for people who wanted to indulge their nostalgia. Yet the customers who frequented these teahouses to appreciate the "drinkable antique" were relatively young people, mostly in their twenties, thirties, or forties. Nostalgia has become "a hot commodity" and "cultural discourse" in metropolitan China since the mid-1990s (Hsing You-Tien 2006: 478). In this context, the transformed teahouses and transformed tea became a pair, catering to the demands of the new generation of consumers for something both old and new. This is consistent with the ongoing consumer revolution in China, which not only changed the substance that was consumed but also "nurtured individual desires and social networks" (Davis 2000: i).

When Lao Li described his past experiences, Zongming was keen to see the traditional teahouses that had a storytelling component. Lao Li, Hongtu, and I took him for a walk along the central axis of Kunming City, where only small parts of the old Qing- and Republic-period neighborhoods are left. To Zongming's disappointment, we didn't find any old teahouses. In fact, many of the old houses were now enclosed by walls. It was said that a newer round of protection and restructuring by the local government was beginning. Eventually, we took Zongming to dinner at a restaurant in a redecorated traditional house. The courtyard and the surrounding tile roofs presented a nice quadrangle, like the Chinese character "口" (*kou*, mouth). Sitting in the courtyard of the wooden dwelling, Zongming was served a meal of classic Yunnanese cuisine. All through the meal Puer tea was served. Zongming enjoyed the food but didn't drink much of the tea. He didn't seem to think that Puer tea served in one of these "traditional" tea restaurants tasted good. After the meal, he asked to drink tea back in Hongtu's tea shop, which, although smaller and less luxurious, was more to his liking.

CONCLUSION: MULTIPLE VISIONS

The interplay between the places of Puer tea production and consumption is not new. But never has it been as important as in the current period for consumers to clarify the identity of Puer tea, an endeavor that has made the tea's authenticity more controversial. This controversy is situated in the context of China's transformation, the consumer revolution, and the desire of undeveloped regions, such as Yunnan, to promote themselves.

As John Durham Peters pointed out, modern men and women view the world through two lenses simultaneously: through their own eyes, to see the locality close by, and through the modern media, to see the process of globalization (Peters 1997). Yunnanese people use this "bifocal vision" to view Puer tea not only from the local perspective but also from the perspective of others, namely the Cantonese or the Taiwanese. Like Puer green tea, Yunnan has had to take upon itself more than one identity: following the consumption custom that had long been shaped by the local environment; adopting the aged value of Puer tea from the outside and serving it to the global market; taking advantage of the tea's raw value and making Puer tea a new local representation. Therefore, conflict began, interwoven with multiple voices. And in the endless *jianghu* debate, there is always a counterforce against the temporarily established version. Puer tea's image became complicated. Rather than having a single and localized vision, it had mobile and multiple visions. One distinct taste preference, or one defined authenticity, is never intrinsic and fixed but is always on the way to being updated, moving along with temporal, spatial changes and with changes in the interests of those who seek to define it.

CONCLUSION

An Alternative Authenticity

Puer tea has been packaged into a popular beverage that has attracted the fascination of many people in twenty-first century China. Its packaged values, in turn, have been debated, counterpackaged, and reinterpreted by multiple actors. The packaging process contains several parallel narratives of transformation. The first is the transformation of Puer tea's profile from something common and unnoticed to something extraordinary, valuable, and representative of high culture. The second is the new discourse about the value of aging: in production, older tea trees are regarded as much better than younger tea bushes; in consumption, the tea needs to be "cooked" to transform it from raw to aged, and the longer tea is stored, the higher its value. As a result, forest tea is preferred for cultivation and production, and the flavor of aged tea is preferred by consumers. Along with this change in tea production and consumption, a third transformation occurred in the representation of Yunnan, from a remote, undeveloped, and earthy area into an enchanting land with natural beauty and rich ethnic culture. These transformations have all taken place against an important backdrop—the transformation of China's economy and society, which is exemplified by the reemergence of a national fever for tea culture and other forms of consumption revolution. Within these general transformations, there have been many partial and back-and-forth changes during certain periods and in certain places. The authenticity of Puer tea has developed unevenly and unstably, subject to redefinition by history and context. The unresolved authenticity of Puer tea is part of the changeable social landscape; it embodies the transformation of people's understandings about national, regional, and individual identities.

The counterpackaging process has reinforced the unstable status of Puer tea's authenticity. The unresolved authenticity of Puer tea lies in the multiple counterforces that are present in the unpackaging narratives. Endless debates, controversies, suspicions, and revisions relate to Puer tea's meanings, values, regulations, and representations. One force resists accepting the aged value of Puer tea and endeavors to reexplore and reinterpret its raw value, while another force deconstructs Puer tea's propagandized value and wants to draw it back to its "original" meaning in order to obtain a properly balanced value. Yet another force resists persuasion, guidance, cheating, and authoritative instruction, asking tea drinkers and traders to remain loyal to their own judgment and to solve problems using their own skills. And still another force subtly resists tough regulation, exploring the margin beyond any regulation and relying on itself to transform indigestible standards into acceptable practices.

The above narratives and debates embody what I call the *jianghu* of Puer tea. These *jianghu* forces are particularly evident in the unpackaging narratives that grow out of the popular realm and stand for nonmainstream voices. The chaotic situation of Puer tea illustrates the intrinsic feature of *jianghu* culture. *Jianghu* is filled with risk, suspicion, vagueness, and contention. The debate and unpacking of Puer tea's packaged values also fully embodies the *jianghu* actors' ability to cope with risks, debates, and nonstandardization. These features of *jianghu* culture embody important enduring as well as transformed characteristics of Chinese cultural consumption. This Chinese *jianghu* culture is crucial in helping us understand the chaotic situation of Puer tea today, as well as the responses to this chaos adopted by multiple Puer tea actors.

After experiencing a violent rise and fall in 2007, Puer tea production and consumption continues, but not to the same extent as when "the entire nation was engaged in tea." As many tea friends comment, "Only those who are truly interested in Puer tea remain in this circle." The general tea price has developed stably year by year, more due to the increasing standard of living and price levels in China. The price of forest tea, nevertheless, has increased more noticeably, and the price gap between forest and terrace tea has expanded. In the autumn of 2012, the price of forest tea in Yiwu generally reached ¥600 per kilogram, sometimes as high as ¥1,300, whereas the price of terrace tea remained around ¥85. The value of forest tea has increased steadily, along with middle- and upper-class Chinese consumers' pursuit of luxury goods. But the amount of forest tea is limited, and the competition grows fiercer. Several friends who continue to trade tea in Yiwu report that the *jianghu* in Yiwu has become even riskier.

The *jianghu* concept sheds light on how the Puer tea fad and the debates around it have been shaped by a specific cultural and social context. A broader cultural comparative perspective could be used to deepen this understanding. A good comparison would be with red wine in the West. Both Puer tea and red wine are said to increase in value as they age, and tastes for both are shaped by growing environments, processing techniques, and storage conditions. Of course, the social lives of both tea and wine are influenced by their social and cultural contexts. These are important similarities, but an important difference is that it is far easier to authenticate red wine than Puer tea. The information on a bottle of red wine seems more authoritative to an ordinary consumer in Australia, America, or Europe than the information on a piece of Puer tea in China. The categorization of red wine has been more definitely fixed; many of its flavor descriptions have been documented and widely shared; and most government regulations concerning wine appear to be more standardized than the regulations for Puer tea. The distinctive situation of Puer tea invites drawing on the concept of *jianghu* to explore it.

RETHINKING *JIANGHU* AND MULTIPLICITY

Should my informants read this book one day, I suspect that some would appreciate the *jianghu* analogy because they have both suffered and benefited from the *jianghu* of Puer tea. Some even stated directly that the world of Puer tea is like a *jianghu.* At the same time, many would no doubt be disappointed, because they won't find a singular and authentic narrative in this book about Puer tea. Instead, they may feel that the multiple voices described here will only make more trouble for readers. That is, although many of them regard the world of Puer tea as a *jianghu,* resist the dominant singular voice from the government, and have sorted out flexible ways to survive in this chaotic and vague *jianghu,* in ideal terms they all wished that there were a single clear and authentic definition of Puer tea. During my fieldwork, I met numerous people who showed great interest not only in tasting Puer tea but also in knowing about Puer tea. Many of them commented that previous writings about Puer tea had failed to cover all the facts or had provided false information, even though some of the popular writings used words like "full display" (*daguan*), or "most authentic" (*zui zhenshi*) in their titles. These informants may have hoped that my book would help to resolve some of the uncertainties. They may be disappointed or, even worse, think I am criticizing Puer tea or Yunnan, or that I am providing misleading information, which would negatively impact Puer tea's development in Yunnan.

Imagining these possible responses prompts me to rethink why I have developed my writings within the *jianghu* framework. It also makes me reflect more on the anthropological perspective and methodology in exploring the authentic meanings of Puer tea as well as other consumer goods.

When I started this research, I endeavored to discover authentic facts about Puer tea, at least for the historical sections. But soon after I began my fieldwork, I found it difficult to clarify many issues. Even a question like "What is Puer tea?" had numerous answers. I was swayed by various actors and became confused. It seemed that everything everyone told me was believable as well as unbelievable. I doubted that I could prepare a book about the history of Puer tea, given that a singular and authentic historical truth about it may never be verified.

Many events involving Puer tea were intertwined with complex human negotiations, and it became hard to clearly identify what was authentic and what was not. My initial interest in Yiwu had been partly inspired by Zhang Yi, the retired township leader who pioneered the recovery of handcrafted caked Puer tea in Yiwu in the mid-1990s. Many tourists, traders, and journalists visited him to learn about the local history of Puer tea. I visited him many times, too, and he spared time to answer my questions patiently. He published his own book in 2007 about the history of Puer tea in the Six Great Tea Mountains. He inscribed the copy I bought from him on the blank page at the front with the following words: "A tea person should care very much about honesty and faith" (*Charen yi chengxin wei zhong*). That day he talked to me about how fake Puer tea was increasingly becoming a problem in the market. On later visits, I found that he preferred to talk about how to cultivate tea properly in a healthy ecosystem rather than about historical events. Mr. Zhang said he wished to write another book focused on tea cultivation, but he was worried that he might not be able to do so, because he was old and had not been well in recent years. I agreed that, to a large degree, the quality of the tea plant might be more important than the quality of the historical truth. And I understood that maybe he had been distracted by too many historical debates.

Through gossip, I heard that some people, both locals and outside traders, were suspicious of Mr. Zhang and his tea. They said that some of the historical issues he wrote about in his book were untrue. They said that he provided false information for his own commercial benefit. Some local villagers said it was good that he had taken a lead in boosting the production of Puer tea in Yiwu in recent years, but it was bad that he had taken a lead in pollarding forest tea trees and planting terrace tea fields in the early 1980s. Some traders had bought his earliest tea products but suspected that they had not been made from pure forest tea material, as their flavor had not improved even after almost ten years. I once visited Zhang's tea terraces and saw that he had tried his best to make some adjustments, increasing the gap between the tea trees, almost like forest tea. As for his earliest tea products, he had no way of making any adjustments, as the tea had been sold ten years earlier and stored by somebody else. The suspicions and criticism became fiercer in early 2007, when Puer tea's price reached its peak, as any judgments about older products at this moment could easily influence the value of new products. In these debates, Mr. Zhang was seen as someone who had done both good and bad for Yiwu's Puer tea.

Mr. Zhang died at the end of 2008, at the age of sixty-seven. He had suffered from heart disease for a long while, and some of his neighbors in Yiwu said the suspicions and criticism of him and his tea had made his condition even worse. After his death, many visitors wanted to pay their respects at his grave, but they were rejected by his son, who perhaps didn't want his father to be disturbed any more, whether by respectful praise or negative gossip.

Mr. Zhang's story shows how complex Puer tea's historical "truth" can be. It is difficult, and perhaps also unnecessary or even impossible, to obtain material that is 100 percent authentic either for tea or for history. Both the quality of tea and the quality of history become hard to verify. If one aspect is defined too absolutely, it could easily turn out to be incorrect. These confusions and complexities made me gradually realize the value of multiplicity rather than singular authenticity. The multiple voices about Puer tea debate and complement one another; each voice stands for a certain purpose, contains certain values, and represents certain meanings in relation to the others. Setting out the details about this debate, divergence and interaction provide alternative narratives about Puer tea. This concern has drawn my attention toward those voices emerging from the popular realm, in which multiple values coexist and forces and counterforces complicate existing notions regarding Puer tea's authenticity. It is a *jianghu* arena where multiple actors dispute, interact, and solve problems contextually. In this arena, Puer tea's authenticity cannot be standardized. It can be managed only via interpersonal negotiations. In this regard, I have taken Mr. Zhang's voice to be indispensable among the multiple voices in the *jianghu* of Puer tea, and I have also admitted that he is one of the people who have played an important role in Puer tea's history.

Thus I came to enjoy being an anthropologist, conducting participant observation, acting as an audience member watching the actors' performances, asking questions when necessary, and avoiding judgments about the authentic truth of Puer tea. But, of course, my own understandings about Puer tea were inevitably and contextually influenced by certain informants and certain popular writings. "To immunize against *huyou*" is never possible, and "being loyal to one's own senses" is actually mixed up with conscious and unconscious acceptance or rejection of external influences. Although this book doesn't attempt to present absolute truth about Puer tea, many of its narratives reflect my intentions, displaying agreement or disagreement with different points of view. Does authenticity exist? From the anthropological point of view, any statement or behavior can become abstract, nonemotional, and understood in terms of constructed or packaged meanings. The boundary between the authentic and the fake is blurred, and both are influenced by politics and used for constructing certain identities. In everyday life, however, one does not need to realize that every food and every meal is political.

FILM AND ALTERNATIVE NARRATIVES

Film has been used as an important methodology in the research underlying this book, and it provides alternative narratives and visual information. Specifically, the film segments listed in appendix 2 trace the social life of Puer tea from production to consumption, from harvesting in the tea fields or forest to processing by local families, trading in both rural and urban markets, promotion at various events, and finally tasting by tea drinkers. Bearing in mind the complementary relationship of video and texts helped me make decisions more clearly about when to film and when not to.

Rather than making a single complete film, I have made seven independent shorter films plus two still presentations. They complement one another, and each complements a certain book section, although the films have not been edited to parallel the book exactly. Films have their own rhetoric, their own logic of narration, and their own content beyond the text. Rather than fully dominating the narrative like textual ethnography, film, as the ethnographic filmmaker David MacDougall has pointed out, "to some degree allows one to look over the shoulder of the film-maker, albeit from the position that the film-maker chooses" (Grimshaw and Papastergiadis 1995: 32). In a way, using film echoes the *jianghu* theme adopted in the book, showing how Puer tea is defined in multiple ways. Films extend these multiple narratives from text to video, further displaying the multiplicities in daily life and allowing the audience multiple interpretations.

In chapter 2, we saw the trader Wen's displeasure when he found that tea growers had blended terrace tea with forest tea. In order to keep his business relationship with the growers, he had to back down and ultimately accept some impure tea leaves. This was what I witnessed, filmed, and wrote. But in the corresponding section of film, "Authentic Tea," I edited the story in another way: Wen was unhappy with the blended tea leaves, he scolded the growers, and then he went to a nearby family to buy more tea material. This film segment does not show whether he bought the impure tea leaves. One major reason for this ambiguity is that I felt that the interference of my camera had influenced the unfolding of the event. Wen's decision to buy the mixed tea material was, of course, concerned with preserving a stable business relationship, and it might also have been concerned with saving face for the tea growers in front of the camera. But he might also have worried that if he bought the tea, this "inauthentic" event would be recorded. I now see that it was actually a very difficult moment for him: either buying or not buying would have been a problem. When I saw him encounter impure tea leaves on another occasion, when the camera was not present, he lost his temper and rejected the trade. I came to understand that Wen was a critical tea trader and that, whenever possible, he preferred to buy nonblended tea leaves. So, rather than editing the film of the event to show that he bought the impure tea leaves—this was only one case among many trading experiences he had in Yiwu—I chose to edit it with an unclear ending to imply an alternative version of reality, and left it open for the audience to ponder.

In another film, *Spring Harvest,* my initial intention was to show the procedures of harvesting and processing terrace tea and forest tea. I chose to film the Gao family in Yiwu. Completely unexpectedly, some events related to a family dispute intruded on their tea work. While picking and processing tea, Mrs. Gao and her daughter complained that the daughter-in-law was lazy and didn't help them with any work, and Mrs. Gao expressed her frustrations and anger toward her son as she stir-roasted the tea leaves in the evening. One possibility for editing this film was to ignore their family dispute and show only the picking and processing procedures. But I felt that these events couldn't really be separated from the family's tea work and livelihood. All the hardness, criticism, and emotion was stirred into the tea leaves by Mrs. Gao as she roasted them. Tea is thus far more than a drink to quench thirst. As many of my informants taught me, Puer tea would have no meaning if it were regarded as simply a broth or a liquid; instead, it must be experienced and interpreted along with multiple aspects of social life.

APPENDIX 1

Puer Tea Categories and Production Process

PUER TEA CATEGORIES

Date of Production

Old family commercial brands (Haoji). These are the earliest Puer tea products, produced during the late nineteenth and early twentieth centuries by private family tea companies in what is now Xishuangbanna (including the Six Great Tea Mountains and Menghai) and Simao. Those originating in the Six Great Tea Mountains are regarded as the earliest. Most of this tea is now kept in museums or held by connoisseurs, mostly in rounded cake form. Famous brands include Tongqing Hao, Songpin Hao, Tongxing Hao, and Tongchang Hao, all of which were old family commercial brands in the Six Great Tea Mountains (figs. 1.1 and 1.2).

Puer tea imprinted with the Zhongcha brand (Yinji). Zhongcha is the brand name of the Chinese Tea Company, and for Puer tea it refers specifically to the Yunnan Provincial Branch. The Zhongcha brand mark is composed of the Chinese character *cha* (茶) in the center, encircled by eight *zhong* (中) characters (figs. 1.3 and 4.2). The tea was produced from the late 1930s to the 1980s by the national tea factories of Menghai, Kunming, and Xiaguan.

Modern (Xiandai) tea produced since the 1990s, when private tea companies reemerged and the Puer tea industry boomed in Yunnan.

Processing Techniques (Especially Fermentation)

Raw Tea (sheng cha or sheng pu)

Raw Puer tea hasn't been fermented and is closer to green tea. It can be very astringent when young. Raw Puer tea is available in loose form or made into various compressed shapes.

Artificially Fermented Tea (shu cha or shu pu)

The technique of artificial fermentation was formally invented in Kunming in 1973. By subjecting raw tea leaves to a specific temperature and humidity, the fermentation of Puer tea (mainly microbial enzymatic reaction) is completed within two or three months. Artificially fermented Puer tea is also available either compressed or in loose form.

Aged Tea (lao cha)

This is raw Puer tea that has been "naturally" stored for at least five years, though clear agreement hasn't yet been reached on how many years' storage is required. It is believed that "natural" fermentation (mostly oxidation, possibly also with some microbial enzymatic reaction) occurs during long-term storage. The older the tea, the higher its price. But "natural" is a relative concept, because some people also create a humid storage environment to accelerate fermentation, which resembles the technique used to produce artificially fermented Puer tea. Some artificially fermented Puer tea that has been stored for several years is also considered aged tea.

Outward Appearance

Loose Tea (sancha)

Puer tea in loose form (see fig. I.3).

Compressed Tea (jin cha)

Puer tea in various compressed shapes, including round, brick, mushroom, and bowl-shaped (see fig. I.4).

PRODUCTION PROCESS

Rough Processing and *Maocha*

The process of harvesting, sorting, "killing the green" (see below), rolling, and drying. The final product of rough processing is *maocha*, the dried basic tea leaves in loose form.

Harvesting/Picking

Harvesting starts in February or March. Tea leaves sprout continuously throughout the spring, summer, and autumn (usually until November), but spring tea is the best. Summer tea is regarded as inferior because the abundance of rainy days could result in a lack of aroma and increased astringency in the flavor of the tea. In China, people usually pick one bud

TABLE A.1 An overview of Puer tea categories and brew characteristics

Categorization Standard	Category	Color (Brew)	Aroma (Brew)	Taste (Brew)
Date of production	Haoji	—	—	—
	Yinji	—	—	—
	Xiandai	—	—	—
Processing technique	Raw tea	Faint yellow	Fresh; flower and honey aroma	Astringent; brisk; lingering sweetness after bitterness
	Artificially fermented tea	Strong, bright red	Earthy or moldy; jujube aroma	Smooth; sweet and bitter
	Aged tea	Strong, bright red	Earthy or moldy; flower or fruit aroma	Smooth; mellow; thick; lingering sweetness
Outward appearance	Loose tea	—	—	—
	Compressed tea	—	—	—

Note: it is not possible to provide standard brew characteristics for Haoji, Yinji, and Xiandai tea, or for loose and compressed tea, as each of these categories could be subdivided into raw and fermented versions. Even for raw, artificially fermented, and aged tea, the descriptions provided here are only of the general cases. More nuanced differences exist among them depending on what tea resources, processing techniques, and storage methods are used.

plus two leaves from each sprout, but for Puer tea they often pick two or three extra leaves.

Sorting

Sorting involves removing rotten or fragmented tea leaves and separating the leaves into different grades. Fresh tea leaves are sorted soon after picking, and *maocha* is also sorted for further fine processing.

"Killing the Green" (sha qing) / Stir-Roasting (chao cha)

"Killing the green" is one way to deactivate oxidation and the action of enzymes and to suppress fermentation in tea leaves. Different methods of killing the green result in different flavors of tea. For example, steaming is popularly used on Japanese green tea, and sun-drying was used before other methods were invented. Puer tea leaves are stir-roasted without oil. Traditionally, fresh tea leaves were placed in a large, dry wok heated by charcoal or wood, and workers used gloved hands or bamboo sticks to turn the tea

TABLE A.2 An overview of the Puer tea production process

General Production Procedure	Detailed Production Procedures	
Rough Processing	Harvest—sort—kill the green—roll—dry—raw *maocha*	
Fine Processing	Raw Puer tea	Sort—weigh—shape—compress—dry—wrap
	Artificially fermented Puer tea	Stack and moisten—turn over and ferment—artificially fermented *maocha*—sort—weigh—shape—compress—dry—wrap
Storage and further fermentation		

leaves until their color and quality changed, but in modern processing, stir-roasting is done by a machine.

Rolling

The purposes of rolling are to (1) achieve different tea leaf shapes; (2) facilitate storage, maintaining crispness and avoiding breakage; (3) allow the tea brew to be easily released in later infusing, and, with different degrees of rolling, to result in different flavors. Again, rolling was traditionally done by hand but is now done by a machine.

Drying

Drying occurs at various stages of processing. Loose tea leaves are dried after rolling, and compressed tea is dried before wrapping (and sometimes also after wrapping). In fine weather, tea is dried in the sun; in inclement weather, it is baked (by fire or in a large oven).

Fine Processing

Fine processing turns loose *maocha* into compressed and wrapped tea. In fine processing for raw tea, *maocha* is steamed, shaped by machine or by hand, pressed, dried, and wrapped. In fine processing for artificially fermented Puer tea, *maocha* is piled indoors under a specific temperature and humidity. A microbial enzymatic reaction, one kind of fermentation, takes place to mature the tea. This usually takes two or three months, during which the piled tea material needs to be turned over several times to ensure that it is completely fermented. Then, after drying, the same fine processing procedures used on raw Puer tea are applied: steaming, shaping, compressing, drying again, and wrapping.

Storage

All Puer tea—whether compressed and wrapped or loose *maocha,* raw or artificially fermented—can be put into storage. This is increasingly regarded as an extended and essential part of Puer tea's production, and it is said that the taste of Puer tea is improved by long-term storage.

Fermentation

Puer tea is fermented by two methods. The first is oxidation. When tea comes into contact with air, oxidation happens. As stated above, killing the green suppresses the oxidation to a certain degree. The second method of fermentation is the result of a microbial enzymatic reaction, which is used for artificially fermented Puer tea. Natural fermentation occurs during long-term storage of either raw or artificially fermented Puer tea. "Natural" is a relative concept, as some artificial methods—such as creating a humid storage environment—may be further applied. In this "natural" postfermentation, both oxidation and microbial enzymatic reaction may occur, depending upon the temperature and humidity of the storage.

APPENDIX 2

Supplementary Videos

Segment	Title	Length	Corresponding Text
Film 1	Twice Puer Tea in Hong Kong	15 minutes	Chapter 1
Film 2	Spring Harvest	29 minutes	Chapter 2; conclusion
Film 3	Visiting Yiwu, Tasting History	30 minutes	Chapter 1
Film 4	Spring Tasting	35 minutes	Chapter 2
Film 5	Authentic Tea	24 minutes	Chapter 2; chapter 5; conclusion
Film 6	Walking on Two Legs	23 minutes	Chapter 6
Film 7	Tasting Ancient and Modern	6 minutes	Chapter 7
Presentation 1	Rough Production Process of Puer Tea in Yiwu		Introduction; chapter 2
Presentation 2	Fine Production Process of Puer Tea in Yiwu		Introduction; chapter 2

Note: These videos along with photos used in this book may be accessed at http://dx.doi.org/10.6069/H5WD3XHC.

NOTES

INTRODUCTION

1 I borrow the terms *terrace tea* and *forest tea* from Nick Menzies (2008).

2 According to the *Mengla County Annals* (EBMCA 1994: 226), terrace tea planting started in the 1960s. But in Yiwu, many tea farmers told me that larger-scale terrace tea planting commenced in the 1970s and 1980s.

3 Zhu Zizhen (1996), a scholar of tea history, argues that the transition from compressed tea to loose tea had occurred before this period. But it is likely that the emperor's command encouraged the production of loose tea. Strictly speaking, compressed tea—pressed with loose leaves—is different from molded tea made from well-pounded tea paste and popularly given as a tribute to the emperor during Northern Song (960–1127).

4 The botanical origin of tea is under debate. Some argue that tea originated in India (see Baildon 1877) while others say that China—and specifically Yunnan, with its rich wild tea tree resources—is its birthplace (see Chen Chuan 1984; Chen Xingtan 1994; Evans 1992). Still others integrate various statements, placing its origin in the foothills of the Himalayas or in Southeast Asia, including Yunnan, Burma, Laos, Thailand, and India (see Ukers 1935; Macfarlane and Macfarlane 2003; Mair and Hoh 2009).

5 These areas are famous for producing Puer tea, but the northern, central, and southern areas of Yunnan also have tea resources. In addition to Puer tea, Yunnan also produces green tea and red tea (usually referred to as black tea in English) (Chen Xingtan 1994).

6 Some regard the Hani and Jinuo as the earliest tea harvesters in Yunnan (see Gao Fachang 2009: 25–29).

7 In China, large-leaf tea occurs mainly in Yunnan. Other tea areas, such as Zhejiang, Jiangsu, Fujian, and Sichuan, mainly produce small-leaf tea. Yunnan also has its own small-leaf tea, which was transplanted from Sichuan, to the north of Yunnan.

8 Debates still exist in the tea science about the difference between oxidation and fermentation. In reality, it is often hard to judge when oxidation ends and fermentation begins. I accept the view that oxidation is one type of fermentation (Cai Rongzhang 2006) and thus there are two kinds of fermentation in tea processing: oxidation and microbial enzymatic reaction.

9 Jasmine tea, which has green tea as its base, is the most popular scented tea. Other types of tea may also be scented, such as scented red tea, scented oolong tea, and scented Puer tea.

10 In some areas of China, stir-roasting is still done by hand. Tea farmers use charcoal or wood to heat a large wok; fresh tea leaves are poured into the wok and stir-roasted (without oil), either by hand (wearing gloves) or with bamboo sticks, until their color and intrinsic quality has changed. In modern processing, this is done by a machine.

11 That is the Great Pearl River Delta, which includes Guangdong, Hong Kong, and Macao. In all these regions Cantonese make up the majority of the population.

12 Besides these two opinions, there are also other views, such as that storage was initially undertaken by tea producers in Yunnan who could not sell their tea products and had to keep them for a long period (Ruan Dianrong 2005a: 107).

13 For an earlier phase of the volatile tea industry in Fujian, China, see Gardella (1994).

14 This survey was undertaken by the Promotion Association of Kunming Ethnic Tea Culture and the Long-Run Puer Tea Institution of Yunnan Agricultural University (*Yunnan Daily* 2006c).

15 In fact, several other kinds of dark tea are also compressed, such as those produced in Hunan and Sichuan for export to Tibet and Mongolia.

16 See chapter 63 of *Dao De Jing (Tao Te Ching)*, which "finds flavor in what is flavorless" (Lao Tzu 1998: 132–133).

17 At that time, these stories were called legends (*chuanqi*).

18 However, the martial arts fiction scholar Chen Pingyuan (1997) insists that *jianghu* refers to the space for knights-errant, who are different from bandits and hermits.

19 Written by Shi Nai'an in the fourteenth century, *The Water Margin* is one of the so-called four great classical Chinese novels. Some cite Luo Guanzhong as an additional author.

20 Wu is an independent documentary filmmaker and writer in China. For more information on his life and works, see Filmsea (2003).

21 For the literature on food governnance and regulation in postsocialist countries, see Dunn (2005) and Caldwell (2009).

22 On gift giving and *guanxi*, mainly in rural areas, see also Yan Yunxiang (1996) and Kipnis (1997).

CHAPTER 1

1 Some people call themselves Xiangtang, but they are also officially recognized as Yi.

2 It is not completely clear when the Han became involved in the cultivation of tea. Some records say that the Han joined in tea cultivation in Yiwu soon after their migration, while others recount these events without mentioning an exact date. It seems more likely that, from the time the Han migrated to Yiwu until about 1900, tea cultivation and rough processing were practiced primarily by the indigenous ethnic groups. The Han mostly organized fine processing in commercial companies and participated in trade.

3 For the Western academic accounts about the caravan routes between Yunnan and other Southeast Asian regions, see Forbes (1987); Hill (1998); and Prasertkul (1989). There is also a rich scholarship in Chinese on these routes. See Lei Pingyang (2000); Mu Jihong (1992; 2003); and Liu Qinjin (2005).

4 According to Zhang Yingpei (2006: 5), Li Shi (1988: 1) first notes in his *Xu bo wu zhi* (Sequel of natural history) that the Six Great Tea Mountains had been one of the main tea production areas of Yunnan since the Tang dynasty in the seventh century. There are several versions of the history of the Six Great Tea Mountains: for instance, in some literature, Yiwu is replaced by Mansa, or there is no Mangzhi (Jiang Quan [1980] 2006; Zhao Zhichun 1988).

5 For this history, some of the literature says that the tea materials were sent to Puer, while others say it was sent to Simao. In general, it seems that the tribute tea factory was established in Ninger County under the capital, Puer, for fine processing, and the General Tea Bureau of Simao was responsible for the whole tribute tea task. See Huang Guishu (2005: 88–90); Lei Pingyang (2000: 28); and Ni Tui ([1737] 1981: 593–594).

6 The French built a meter-gauge railway between Kunming and Hải Phòng in 1910, which was then used to transport tea, too.

7 As I have mentioned, many Han inhabitants in the Six Great Tea Mountains were originally from Shiping. Shiping merchants were one of the most famous trade guilds in Yunnan throughout the Ming dynasty, Qing dynasty, and Republican period, and they also had associations in Menghai and Simao (see Luo Qun 2004).

8 Most scholars believe Tibetans started drinking tea in the eighth century. See Hill (1989) and Yang Bin (2004). Yang Haichao (2010) argues that tea drinking dates to the Wei and Jin dynasties. He examines the various names and pronunciations of the word for *tea* in both ancient Chinese and ancient Tibetan throughout different periods.

9 The Tibetan traders usually covered the caravan transport from Tibet to Yongsheng (Lijiang), or sometimes as far as Puer, while the remaining routes were dominated by the Han or Muslim Chinese (Hill 1989). The journeys of Tibetans directly to Yiwu, as mentioned in the following special case, were actually uncommon.

10 This is the general situation. At present a few tea-processing units in Menghai also craft tea by hand, and in Yiwu, there are also some so called tea factories, but these "factories" are smaller in scale than those in Menghai.

11 There were a few families in Yiwu who made artificially fermented Puer tea at their customers' request, but raw Puer tea predominates there.

12 Cooking is often related to artificial work, but ripening is regarded as a more natural process. Mayfair Yang may have mixed them up. What she refers to in her discussion of the gift economy is more applicable to cooking. In this book, I am also talking about cooking, and I stress that this is a transformative process, in which outside traders seek to build up good relationships with local people within a short period of time.

13 See EBMCA (1994); Jiang Quan ([1980] 2006); Li Fuyi (1984); Ruan Fu ([1825] 1981); Tan Cui ([1799] 1981); and Zhao Chunzhou and Zhang Shungao (1988).

14 According to Arjun Appadurai (1996: 49), in the new global order, imagination is becoming "central to all forms of agency." Picking up on the same theme, Marilyn Ivy (1995) traces how the rural *Tōno* is imagined as a nostalgic hometown representation of authentic Japanese tradition. Beth E. Notar (2006b), in her research about Dali in Yunnan, notes that popular narratives become standards of authenticity for visitors in imagining what Dali should be.

CHAPTER 2

1 The Xishuangbanna Supervision Bureau of Technology and Quality wrote a first draft in 2005, upon which the later provincial standard was based.

2 The data from local government indicated that there were fifty tea-processing units in Yiwu in 2007, but I was told by many locals that there were actually around eighty.

3 The Xishuangbanna Supervision Bureau of Technology and Quality has set some guidelines for rough processing, but these guidelines are advisory, not compulsory.

4 In Wen's words, "Its blandness is unique; its aftertaste lingers in the mind" (*dan ding tianxia, huiwei wuqiong*).

5 According to Wen, unpruned tea trees grow more slowly and therefore develop richer flavors.

6 Other research has also provided cases to show that tourism is increasingly viewed as a "destroyer" of culture. See, for example, Oakes (1997) and Hillman (2003).

7 This process parallels the story of milk in America, where everything true in the "progress story" became untrue in the "downfall story" (Dupuis 2002). In the former, milk is imbued with various positive virtues and related to the pure pastoral landscape, but in the latter story many aspects of its quality came under suspicion and significant doubts emerged about whether or not to drink it. The story of milk has unfolded over the past century and a half, whereas the packaging and unpackaging of Puer tea has occurred in only about five years.

8 Notably, these counterfeits were mostly handmade, but they were often hard to distinguish from the originals. However, while talking about the nineteenth-century industrial revolution in the West, Walter Benjamin ([1936] 1999) notes that it was easier to identify a manual copy from the original.

9 For discussions on the difference between Western and Chinese understandings about individualism, and about the rise of the Chinese individual, see Yan Yunxiang (2009).

10 In the late Ming period (the late sixteenth to early seventh century), individualism was embodied by trademarks that were popularly used for many goods in order to protect intellectual property (Clunas 1991: 66–67). And in contrast to the Maoist era, when unification was demanded, the Reform era allowed more space for variation and self-presentation. Following the Western model, new individual identities came on the scene, often embodied in the freedom of consumption choices (Croll 2006).

CHAPTER 3

1 This is a process that has also happened elsewhere—for example, in Shangri-la, as mentioned by Ben Hillman (2003) and also described by Tim Oakes and Louisa Schein (2006).

2 To learn more about the detailed administrative changes of this area, see *Pu-Erh* (2007b) and EBSAA (1996).

3 This tea is called either *jin gua gong cha*, as it looks like a golden melon, or *rentou gong cha*, as it also looks like a human head, or together as *rentou jin gua gong cha*, Human Head–Golden Melon Tribute tea.

4 The transliterated term is *Puer,* according to the Chinese pinyin spelling system, or *Puerh,* according to the Wade-Giles system used by Western scholars in the past.

5 This was initially stated by Ruan Fu during the Qing dynasty ([1825] 1981: 396) and later cited in many other works. See Li Fuyi ([1939] 2000: 57) and Zhang Shungao (1988: 79).

6 The Chinese novel *Dream of Red Mansions* (Honglou meng), written by Cao Xueqin during the eighteenth century, includes a scene describing how the concubine Jia Yuanfei, a daughter from the Jia family, went back home for a short visit, kindly approved by the emperor.

7 Some other tea experts who had participated in tasting the Golden Melon claimed, alternatively, that such aged Puer tea had become tasteless. See Deng Shihai (2004: 15).

8 Also known as the Water Festival, the Dai New Year was a New Year's celebration held in Yunnan as well as in many Southeast Asian countries, including Thailand, Laos, Burma, and Cambodia.

9 This caravan was jointly organized by the Youth Foundation of Yunnan, the Tea Association of Yunnan, and the Commercial Club of Yunnan. According to media reports, all the revenue from the auction was donated to the Hope Project of China, a foundation that patronizes education in undeveloped areas of China (Riftea 2007).

10 The eleven subareas are Kunming, Chuxiong, Yuxi, Honghe, Wenshan, Simao, Xishuangbanna, Dali, Baoshan, Dehong, and Lincang.

11 However, this classification approach also admits that tea material from a certain place can be more suitable for producing a certain kind of tea. For instance, it is acknowledged that the medium-sized tea leaves from Fujian or Taiwan are better than the large tea leaves of Yunnan for making oolong tea.

12 For Western academic accounts about the caravan routes between Yunnan and other southeast Asian regions, see Forbes (1987); Prasertkul (1989); and Hill (1998).

13 For films made in Yunnan, especially the fiction films since 1949, see Li Miao (2006).

14 This point is debated; according to some researchers, other important goods—such as salt, cotton, opium, and guns and other metal tools—were carried (see Prasertkul 1989; Hill 1998; and Giersch 2006).

15 The Swedish East India Company opened up sailing and trade routes to Asia in the

eighteenth century. *Götheborg* was one of the biggest boats in the fleet but sank in 1745 near Sweden after its third voyage to Guangzhou, China. Salvage operations were conducted in the following century and proved that Chinese tea, porcelain, and silk were the main transported goods of that time. The Swedish East India Company started to rebuild the *Götheborg* in 1995, and the rebuilt vessel sailed towards Guangzhou, China, again in 2005 (*Yunnan Daily* 2006b).

16 For a more detailed discussion of *suzhi*, see Andrew Kipnis (2006).

17 Because raw Puer tea neither requires artificial fermentation nor has to be naturally fermented as long as aged Puer tea, it can be produced more quickly.

CHAPTER 4

1 The number 75 refers to a certain way of blending tea leaves initially applied in 1975; the third digit, 7 or 4, refers to the grade of the basic tea material; and 2 stands for the Menghai Tea Factory. Dayi 7572 is artificially fermented tea, and Dayi 7542 is (naturally fermented) raw tea. Both of them are products of the Menghai Tea Factory.

2 This was a modification of the popular slogan "The entire nation was in arms" (*quan min jie bing*).

3 Here *bankers* refers to powerful Puer tea investors, including some big tea companies.

4 The fees are not clearly specified in the formal notification. Some traders I spoke with told me that the geographical mark for each Puer tea cake (357 or 400 grams) cost ¥0.09 (for members) or ¥0.13 (for nonmembers).

5 I heard from many tea traders that this didn't really assure the authentic quality but was just a new way for the local government to collect more fees.

CHAPTER 5

1 Judith Farquhar examines the politics of food in China through three works of popular literature, all touching upon eating: *The White-haired Girl* was written in the 1940s; *Hibiscus Town* and *The Gourmet* were both published in the 1980s.

2 As an anthropologist as well as a practitioner in Chinese medicine, Judith Farquhar (2002) uses the holistic way of treating illness in the individual body as a metaphor for curing diseases in the national body. She argues that "everyday life in Reform China is still inhabited by the nations' Maoist past" (Farquhar 2002: 10).

3 In the early 1990s, a collectively owned tea factory was established in Yiwu and locally produced *maocha* was sent to it as well.

4 Some local people told me that in the 1960s terrace tea planting had also been advocated by the local government, but not on the scale of the early 1980s. This information was echoed in the Mengla County Annals (EBMCA 1994: 226).

5 To pollard a tree is to cut off its top and branches.

6 I interviewed traders from both Hong Kong and Taiwan. Generally speaking, it is acknowledged that the forest tea preference was initiated by the Taiwanese. But the habit and concept of storing and drinking aged Puer tea was initiated by Hong Kong people, and later co-opted by the Taiwanese.

CHAPTER 6

1 Like most areas of Xishuangbanna and nearby Southeast Asia, Yiwu has dry and wet seasons. The dry season lasts from October to May, and the balance of the year is the wet season. The annual rainfall in Yiwu is about 1,500 to 1,900 millimeters, and the wet season accounts for 80 percent of the annual rainfall (EBMCA 1994: 55; Zhao Rubi 2006: 4).

2 See the talk between the Master and his student in *The Analects* (chapter 12). The Master says that after people become "multiplied" and are "enriched," the next step is to "instruct them" (Confucius 1997: 162–163).

CHAPTER 7

1 The website can be found at http://www.sanzui.com.

2 For other aspects of research that stress that identity is constructed via relations with others, see Ohnuki-Tierney (1993) and Notar (2006b).

3 However, by early 2009, when I returned to Kunming, these events had become infrequent as a result of the continuing recession in the Puer tea market.

4 These interpretations have been drawn from insights from the Baidu website: http://zhidao.baidu.com/question/17288829 (accessed 20 June 2010).

5 Once the tea is put in the pot it is brewed several times, depending upon the category and amount of tea. I refer to each of these brews as a "run."

6 In the English version of Jin Yong's *The Book and the Sword*, the definition of "inner force" given in the glossary is: "the part of *kungfu* concerned not so much with particular techniques (moves, styles), but with the basic underlying physical (breathing, posture, etc.) and spiritual (meditation, concentration, consciousness) training, which gives the techniques their inner strength" (May and Minford 2004: xix).

7 Shaolin *kungfu* is a martial arts style named after the Shaolin Temple in central China. It is often practiced with sticks and vigorous motions. By contrast, the Taiji (Grand Ultimate) school of *kungfu* stresses inner spiritual training. For further details, see the glossary in the English translation of Jin Yong's *The Deer and the Cauldron* (Minford 1997).

CHAPTER 8

1 There are many works dealing with how localization copes with globalization in consumption. See, for example, Watson (1997); Wu and Tan (2001); Wu and Cheung (2002); Grasseni (2003); and Dikötter (2007).

2 Mengku is part of the Autonomous Lahu, Wa, Bulang, and Dai County of Shuangjiang in Lincang.

3 The annual average humidity in Kunming is 71 percent while in both Guangzhou and Hong Kong it is 78 percent. This information is drawn from the Baidu website: http://wenku.baidu.com/view/43105e6d011ca300a6c39022.html (accessed 10 April 2009).

4 It may seem strange that Zongming regards raw Puer tea as "cold" and therefore harmful but likes to take a cold shower every evening. However, the coldness of the shower isn't the same as the coldness of raw Puer tea. The former refers to the temperature of the water. Many Hong Kong people take cold showers because the weather is humid and mostly warm. The latter, by contrast, refers to the intrinsic quality of the tea in terms of Chinese medicine. For example, the intrinsic quality of lychee is taken to be hot; even after being frozen, a lychee is still considered hot.

5 Some researchers argue that artificial fermentation techniques such as these had been utilized by the Yunnanese much earlier, since at least the 1930s (Yang Haichao 2007).

6 Dian is the abbreviation for Yunnan, and *qing* literally means "blue-green." Dian Qing was juxtaposed with Dian Lü (*lü* means "green"), Yunnan's other green tea, which is dried mostly by baking.

7 The nine wholesale tea markets were Kangle, Jinshi, Xiongda, Tangzixiang, Qianwei, Juhua, Xiyuan, Dashanghui, and Jinli.

8 This survey was conducted by the Promotion Association of Kunming Ethnic Tea Culture and the Long-Run Puer Tea Institution of Yunnan Agricultural University (*Yunnan Daily* 2006c).

9 Many private tea companies in Hong Kong supply tea for local tea restaurants or international traders, but they don't necessarily open tea shops or teahouses.

10 It is uncertain in what year the survey was conducted.

GLOSSARY

Anhua *Hei cha* 安化黑茶	Anhua dark tea
bai ban 白版	blank version
bai cha 白茶	white tea
Bai Mudan 白牡丹	White Peony
Baihao Wulong 白毫烏龍	White Hair Oolong
baihua 白話	exaggerated or nonsense words
bei qiu 悲秋	lament for autumn
ben ren 本人	the indigenous people
bing cha / yuan cha 餅茶/圓茶	round cake (shaped) tea
boyin chaguan 播音茶館	broadcasting teahouses
bu hexie 不和諧	inharmonious
bu qi 補氣	supplementing the vital breath
Cha chan yi wei. 茶禪一味。	Tea and Chan Buddhism have the same flavor.
Cha Jing 茶經	*The Classic of Tea*
Cha Ma Gu Dao 茶馬古道	the Ancient Tea-House Road
cha qi / ba qi 茶氣/霸氣	tea energy
Charen yi chengxin wei zhong. 茶人以誠信爲重。	A tea person should care very much about honesty and faith.
cha yi shi 茶藝師	the craftsmanship in tea infusing
chaboshi 茶博士	tea doctor
chachang 茶廠	tea factory
chadui 茶隊	tea team
chaguan 茶館	teahouse
chahui 茶會	tea meeting
chao cha 炒茶	stir-roasting (tea leaves)
"chao cha" "炒茶"	tea speculation

chao gu 炒股	speculate on the stock market
chao qing 炒青	stir-roasted green tea
chaosheng 朝聖	pilgrimage
chazhuang 茶莊	household tea unit
chen yun 陳韻	aged appeal
chenhua 陳化	aging, fermentation
Cheshun Hao 車順號	Cheshun brand
chuanqi 傳奇	legends
chun sheng 春生	sprouting in spring
chuantong jiating shougong zhizuo 傳統家庭手工製做	handmade traditionally by a family
da shu cha 大樹茶	forest tea
daguan 大觀	full display
Dan ding tianxia, huiwei wuqiong. 淡定天下，回味無窮。	Its blandness is unique; its aftertaste lingers in the mind.
Dian Hong 滇紅	Yunnan Red tea
Dian Lü 滇綠	Yunnan Baked Green tea
Dian Qing 滇青	Yunnan Sun-dried Green tea
dong cang 冬藏	storing and hiding in winter
dou cha 斗茶	tea competition / tea game
Duihua 對話	*Dialogue* (title of a CCTV program)
fa shui 發水	ferment with added water
fan 飯	cooked rice
fengkuang de Puer cha 瘋狂的普洱茶	crazy Puer tea
fu yong feng ya 附庸風雅	cultural pose
fugu 復古	revive the past
gai tu gui liu 改土歸流	replacing *tusi* [native officials] with imperial officials
ganma 乾媽	godmother
gongfu 工夫	a practice that takes great effort
gongfu cha fa 工夫茶法	the craft of tea infusing

guanxi 關係	the network of relationships in Chinese society / socializing through gift exchange
Haoji 號級	old family commercial tea brands
Haoshi bu guo san. 好事不過三。	Good things don't happen more than three times.
he 和	harmony
hei cha 黑茶	dark tea
hong cha 紅茶	red tea (known as black tea in English)
hong nong ming liang 紅濃明亮	strong and bright red
hong qing 烘青	baked green tea
Hong Yin 紅印	Red Mark tea
Honglou meng 紅樓夢	*Dream of Red Mansions*
Hongtu Lantian 紅土藍天	red earth and blue sky
hua 化	transformation
hua 滑	smooth
hua gange wei yubo 化干戈爲玉帛	turning hostility into friendship
huang cha 黃茶	yellow tea
huigan 回甘	long-lasting sweetness in the throat after the strong and bitter flavor (lingering sweetness after bitterness or astringency)
huyou 忽悠	(original) irrelevant, exaggerated, or nonsense words; (paraphrased) any behavior related to cheating, swaying, hyping, wheedling, or agitating someone with fictitious content
jian 件	a trade unit for Puer tea, usually one *jian* has eighty-four cakes of Puer tea in a bamboo basket or a cardboard box, weighing a total of 31.5 kilograms
jianghu 江湖	lit., "rivers and lakes"
Jianghu zai nial? Jianghu zai renxin. 江湖在哪裡？江湖在人心。	Where is *jianghu* located? In the human heart.
jiaohua 教化	moralize, domesticate, or civilize
jin gua gong cha 金瓜貢茶	Golden Melon Tribute tea
Junshan Yinzhen 君山銀針	Jun Mountain Silver Needle tea

kou 口	mouth
kungfu (gongfu) 功夫	martial arts
lao cha 老茶	aged (Puer) tea
Lao Huang Pian 老黃片	Yellow Leaf Puer tea
lao jianghu 老江湖	old *jianghu*
lao sheng cha 老生茶	aged raw Puer tea
leng 冷	cold
lipin cha 禮品茶	gift tea
Liu Da Cha Shan 六大茶山	the Six Great Tea Mountains
Liubao cha 六堡茶	Liubao tea
Longjing cha 龍井茶	Dragon Well tea
lou fo tong (lao huo tang) 老火湯	slowly stewed soup
lü cha 綠茶	green tea
Lü Yin 綠印	Green Seal
Lu Yu peng cha tu 陸羽烹茶圖	*Lu Yu Brewing Tea* (painting)
man man pin 慢慢品	taste slowly
maocha 毛茶	the dried basic tea leaves in loose form, the result of rough processing
Maojian 毛尖	a sort of green tea mostly composed of tea buds
men huang 悶黃	sealed yellowing
Mengding Huangya 蒙頂黃芽	Meng Mountain Yellow Buds
meng cha 猛茶	vigorous tea
menke 門客	retainers originating in eighth century B.C.E. China, who were accommodated by the master and served him
mi 米	raw rice
mi tang 米湯	soup boiled with rice
mianzi 面子	face
minjian 民間	popular realm
Nametian 納麼田	a subvillage in Yiwu township
neigong 內功	inner force
nuan 暖	warm

pin cha 品茶	tea tasting
ping jin shi zao 平矜釋躁	your arrogance is eliminated; your impatience is removed
po si jiu 破四舊	Destruction of the Four Olds
Puer cha 普洱茶	Puer tea
Puer lü cha chun rui 普洱綠茶春蕊	Puer green tea made from spring buds
Qi Hong 祁紅	Keemun Red tea
qi zi bing 七子餅	seven-son tea cake
Qiang da chu tou niao. 槍打出頭鳥。	The bird that stands out is easily shot.
qing cha 青茶	blue-green tea
qing chang chaguan 清唱茶館	pure singing teahouses
qing yin chaguan 清飲茶館	teahouses for pure tea drinking
qiu shou 秋收	harvesting as well as withdrawal in autumn
quan min jie bing 全民皆兵	the entire nation was in arms
quan min jie cha 全民皆茶	the entire nation was engaged in tea
que liang hu 缺糧戶	grain-deficient household
Ren zai jianghu, shen bu you ji. 人在江湖，身不由己。	Since he is in *jianghu,* he could do nothing but follow the law of *jianghu.*
rentou gong cha 人頭貢茶	Human Head tribute tea
rentou jin gua gong cha 人頭金瓜貢茶	Human Head–Golden Melon tribute tea
Rui gong tian chao 瑞貢天朝	Tribute to the Emperor
sancha 散茶	loose tea
Sanzui 三醉	the name of a tea website in China
sha qing 殺青	killing the green
sha zhu fan 殺豬飯	rural family banquets
shai qing 曬青	sun-dried (green tea)
shai qing maocha 曬青毛茶	sun-dried basic tea material
Shaolin 少林	the name of a Buddhist temple in central China, as well as the school of martial arts related to the temple
she di ming quan 舌底鳴泉	the bubbling-up of spring water from the bottom of the tongue

Shen Nong 神農	Divine Farmer
sheng cha / sheng pu 生茶/生普	raw (Puer) tea
sheng cha pai 生茶派	raw Puer tea group
Shi se xing ye. 食色性也。	Appetite for food and sex is human nature.
shou cha 收茶	collect tea
Shou zhong you liang, xin li bu huang. 手中有糧，心裡不慌。	You will not feel panic if you have rice in your hands.
shu cha / shu pu 熟茶/熟普	artificially fermented (Puer) tea
shu cha pai 熟茶派	artificially fermented Puer tea group
Shu da zhao feng 樹大招風。	Tall trees catch much wind
Shuihu zhuan 水滸傳	*The Water Margin*
shuitu 水土	local water and soil
shuitu bu fu 水土不服	water and earth not fitting
shuo shu chaguan 說書茶館	storytelling teahouses
Songpin hao 宋聘號	Songpin brand
Sui you ren zuo, wan zi tian kai. 雖由人作，宛自天開。	It is man-made, but it looks like it was created by nature.
suzhi 素質	personal quality
tai di cha / xiao shu cha 台地茶/小樹茶	terrace tea
Tieguanyin 鐵觀音	Iron Goddess of Mercy tea
tong 筒	stack
Tongchang hao 同昌號	Tongchang brand
Tongqing hao 同慶號	Tongqing brand
Tongxing hao 同興號	Tongxing brand
tu 土	earthy
tu cha 土茶	earthy tea; indigenous tea
tu te chan 土特產	indigenous product
tuo cha 沱茶	bowl (shaped) tea
waigong 外功	outer force
wen zhi bin bin 文質彬彬	elegant and refined in manner
wenhua 文化	culture

wenhua re 文化熱	culture fever
wu wei zhi wei 無味之味	flavorless flavor
Wuyi Yancha 武夷岩茶	Wuyi Rock tea
xia zhang 夏長	growing in summer
Xiaguan Yi Tuo 下關乙沱	Xiaguan Bowl Tea Grade B
xiake 俠客	Chinese knights-errant / wandering swordsmen
Xiangtang 香堂	a self-identified name of an ethnic group in Yunnan, officially recognized as Yi
xiao yu 小魚	a small fish
Xibu da kaifa 西部大開發	Opening the West
Xinshengdai 新生代	*New Generations*
yangsheng 養生	nourish life
yi liang wei gang 以糧爲綱	planting food as a guiding principle
yi qing yue xing 怡情悅性	your mood is elevated; your temper is softened
yi tong jianghu 一統江湖	unifying the *jianghu*
Yinji 印級	Puer tea imprinted with the Zhongcha brand
Yinzhen Baihao 銀針白毫	Silver Needle White Hair tea
Yiwu zheng shan 易武正山	the authentic tea mountain Yiwu
Yuanfei *xingqing* 元妃省親	Homecoming by Yuanfei
yue chen yue xiang 越陳越香	the longer the storage, the better the taste (the longer the better)
Yunnan *xianxiang* 雲南現象	Yunnan phenomenon
Yunnan yingxiang 雲南映像/影響	*Dynamic Yunnan*
zheng qing 蒸青	steamed (green tea)
Zhengmei 蒸酶	Yunnan Steamed Enzyme tea
Zhongcha 中茶	the brand name of the Chinese Tea Company; for Puer tea it refers in particular to the Yunnan Provincial Branch
Zhongcha Huang Yin 中茶黃印	China Tea Yellow Seal
Zhongguo da jia di yi ren 中國打假第一人	the pioneer of cracking down on counterfeits in China
zhuan cha 磚茶	brick (shaped) tea

ziben zhuyi weiba 資本主義尾巴	negative capitalism
zishahu 紫砂壺	purple clay pot
Zong Cha Dian 總茶店	the General Tea Bureau
zui zhenshi 最真實	most authentic
zuo cha 做茶	refine tea

REFERENCES

Allen, Michael W., Richa Gupta, and Arnaud Monnier. 2008. The interactive effect of cultural symbols and human values on taste evaluation. *Journal of Consumer Research* 35 (2): 294–308.

Anderson, Eugene N. 1980. Heating and cooling foods in Hong Kong and Taiwan. *Social Science Information* 19 (2): 237–268.

———. 1988. *The Food of China.* New Haven: Yale University Press.

Appadurai, Arjun. 1986. Introduction: commodities and the politics of value. In *The Social Life of Things: Commodities in Cultural Perspective,* edited by Arjun Appadurai, 3–63. Cambridge: Cambridge University Press.

———. 1988. How to make a national cuisine: cookbooks in contemporary India. *Comparative Studies in Society and History* 30 (1): 3–24.

———. 1996. *Modernity at Large: Cultural Dimensions of Globalization.* Minneapolis: University of Minnesota Press.

Ashkenazi, Michael, and Jeanne Jacob. 2000. *The Essence of Japanese Cuisine: An Essay on Food and Culture.* Philadelphia: University of Pennsylvania Press.

Baidu. 2006. Yunnan mabang jin jing (Yunnan caravans went to Beijing). *Baidu Website.* http://tieba.baidu.com/f?kz=99422486 (accessed 10 December 2006).

Baildon, Samuel. 1877. *Tea in Assam.* Calcutta: W. Newman and Co. of Calcutta.

Barham, Elizabeth. 2003. Translating terroir: the global challenge of French AOC labeling. *Journal of Rural Studies* 19 (1): 127–138.

Baumann, Gerd. 1992. Ritual implicates 'others': rereading Durkheim in a plural society. In *Understanding Rituals,* edited by Daniel de Coppet, 96–116. London: Routledge.

Belasco, Warren, and Roger Horowitz, eds. 2009. *Food Chains: From Farmyard to Shopping Cart.* Philadelphia: University of Pennsylvania Press.

Belasco, Warren, and Philip Scranton. 2002. *Food Nations: Selling Taste in Consumer Societies.* New York: Routledge.

Benjamin, Walter. (1936) 1999. The work of art in the age of mechanical reproduction. In *Visual Culture: The Reader,* edited by Jessica Evans and Stuart Hall, 72–79. London: Sage.

Benn, James A. 2005. Buddhism, alcohol, and tea in medieval China. In *Of Tripod and Palate: Food, Politics and Religion in Traditional China,* edited by Roel Sterckx, 213–236. New York: Palgrave Macmillan.

Bloch, Maurice. 1998. *How We Think They Think.* Boulder, Colo.: Westview Press.

Bourdieu, Pierre. 1984. *Distinction: A Social Critique of the Judgement of Taste.* Translated by Richard Nice. London: Routledge and Kegan Paul.

———. 1989. Social space and symbolic power. *Sociological Theory* 7: 14–25.

Brook, Timothy. 1998. *The Confusions of Pleasure: Commerce and Culture in Ming China.* Berkeley: University of California Press.

Bu Jing An. 2007. *Cha Sheng Yunnan* (Tea grows in Yunnan). Beijing: Jincheng Chubanshe.

CADN (China Administrative Division Net). 2010a. Yunnan Administrative Division. *China Administrative Division Net.* http://www.chinaquhua.cn/yunnan/ (accessed 12 August 2010).

———. 2010b. Yunnan Xishuangbanna. *China Administrative Division Net.* http://www.chinaquhua.cn/yunnan/xishuangbanna.html (accessed 12 August 2010).

Cai Rongzhang. 2006. *Chadao Rumen San Pian: Zhi Cha, Shi Cha, Pao Chao* (Three procedures to the way of tea: Producing, recognizing, and infusing). Beijing: Zhonghua Shuju.

Caldwell, Melissa L., ed. 2009. *Food and Everyday Life in the Postsocialist World.* Bloomington: Indiana University Press.

Caplan, Pat. 1997. *Food, Health, and Identity.* London: Routledge.

CCTV (China Central Television). 2007a. Yunnan Puer fasheng 6.4 ji dizhen (6.4 Richter-scale earthquake occurred in Puer). *China Network Video Report,* 3 June. http://news.cctv.com/special/peearthquake/index.shtml (accessed 1 July 2007).

———. 2007b. Fengkuang de Puer cha (A crazy Puer tea). *China Commercial Report,* 8 January. http://www.cctv.com/program/cbn/20070109/102457.shtml (accessed 4 April 2008).

———. 2007c. Puer cha de paomo po le (The bubble of Puer tea is broken). *Economy 30 Minutes,* 15 June. http://www.shopyn.com/Article/4536.html (accessed 5 March 2008).

———. 2007d. Chahuahui hua cha (Meet for tea and talk about tea). *Xiaocui's Talk,* 23 December, http://news.cctv.com/society/20071223/101947.shtml (accessed 2 January 2008).

———. 2008. Jie du Puer (Decoding Puer tea). *Duihua* (Dialogue), 20 January. http://www.cctv.com/video/duihua/2008/01/duihua_300_20080121_2.shtml (accessed 9 March 2008).

Chan, Kam Pong. 2008. *A Glossary of Chinese Puerh Tea.* Taipei: WuShing Books Publications.

Chen Chuan. (1979) 1999. Chaye fenlei (Tea categorization). In *Zhi Cha Xue* (The science of tea processing), edited by Anhui Agricultural Institution, 14–24. Beijing: Zhongguo Nongye Chubanshe.

———. 1984. *Chaye Tong Shi* (The whole history of tea). Beijing: Nongye Chubanshe.

Chen Jie. 2009. Puer cha si da jiazhi zhi yi: dili jiazhi (One of the four great values of Puer tea: geographical value). *Guo-Zhengkai's Blog.* http://guo-zhengkai.blog.163.com/blog/static/6964896520099124248557/ (accessed 12 December 2009).

Chen Pingyuan. 1997. Jianghu yu xiake (Swindlers and knights-errant). In *Chen Pingyuan Zi Xuan Ji* (A self-anthology by Chen Pingyuan), edited by Chen Pingyuan, 158–184. Guilin: Guangxi Shifan Daxue Chubanshe.

———. 2002. *Qiangu Wenren Xiake Meng, Chatu zhencang ben* (The literati's old dream of being like a knight-errant, with a collection of illustrations). Beijing: Xinshijie Chubanshe.

Chen Xingtan, ed. 1994. *Cha Shu Yuan Chan Di: Yunnan* (The original locality of tea plants: Yunnan). Kunming: Yunnan Renmin Chubanshe.

Chen Zhenqiong. 2004. Chaguan yu Kunming shehui (Teahouses and Kunming society). In *Minguo Shiqi Shehui Diaocha Cong Bian: Zongjiao Minsu Juan* (The social survey series in the republic of China: Religion and folk volume), edited by Li Wenhai, 465–557. Fuzhou: Fujian Jiaoyu Chubanshe.

Chen Zugui, and Zhu Zizhen, eds. 1981. *Zhongguo Chaye Lishi Ziliao Xuanji* (Selected historical documents on Chinese tea), *Zhongguo Nong Shi Zhuanti Ziliao Huibian* (Special collection on Chinese agriculture). Beijing: Nongye Chubanshe.

Cheung, Sidney C. H. 2005. Consuming 'low' cuisine after Hong Kong's handover: village banquets and private kitchens. *Asian Studies Review* 29: 259–273.

Chi Zongxian. 2005. *Puer Cha* (Puer tea). Beijing: Zhongguo Youyi Chubanshe.

Clunas, Craig. 1991. *Superfluous Things: Material Culture and Social Status in Early Modern China.* Cambridge: Polity Press.

Colquhoun, Archibald R. 1900. *The 'Overland' to China.* London: Harper and Brothers.

Confucius. 1981. Lun Yu (The analects). In *Gudai Hanyu Di Yi Ce* (Ancient Chinese, vol. 1), edited by Wang Li, 176–203. Beijing: Zhonghua Shuju.

———. 1997. *The Analects.* Translated by Arthur Waley. Beijing: Waiyu Jiaoxue yu Yanjiu Chubanshe.

Counihan, Carole. 1999. *The Anthropology of Food and Body: Gender, Meaning, and Power.* New York: Routledge.

Counihan, Carole, and Steven L. Kaplan. 1998. *Food and Gender: Identity and Power.* Amsterdam: Harwood Academic Publishers.

Counihan, Carole, and Penny Van Esterik, eds. 2008. *Food and Culture: A Reader.* New York: Routledge.

Croll, Elisabeth J. 2006. Conjuring goods, identities and cultures. In *Consuming China: Approaches to Cultural Change in Contemporary China,* edited by Kevin Latham, Stuart Thompson, and Jakob Klein, 22–41. London: Routledge.

D'Abbs, Peter. 2009. Chinese tea culture as practice and representation: signposts in a changing landscape. Paper presented at the Society for East Asian Anthropology (SEAA) Conference, Academia Sinica, Taipei, 2–5 July.

Dao Yongming. 1983. Yiwu xian Hanzu he xiongdi minzu de jiaowang (The association of Han and other ethnic groups in Yiwu). In *Daizu Shehui Lishi Diaocha, Xishuangbanna Zhi Yi* (Investigations of Xishuangbanna Dai Society, vol. 1), edited by Yunnan Compilation Committee, 60–63. Kunming: Yunnan Minzu Chubanshe.

Davis, Deborah. 2000. *The Consumer Revolution in Urban China.* Berkeley: University of California Press.

Deng Shihai. 2004. *Puer Cha* (Puer tea). Kunming: Yunnan Keji Chubanshe.

Dikötter, Frank 2007. *Things Modern: Material Culture and Everyday Life in China.* London: Hurst and Company.

Dilley, Roy. 2004. The visibility and invisibility of production among Senegalese craftsmen. *Journal of the Royal Anthropological Institute* 10 (4): 797–813.

Dunn, Elizabeth C. 2005. Standards and person-making in East Central Europe. In *Global Assemblages: Technology, Politics, and Ethics as Anthropological Problems,* edited by Aihwa Ong and Stephen J. Collier, 173–193. London: Blackwell.

DuPuis, E. Melanie. 2002. *Nature's Perfect Food: How Milk Became America's Drink.* New York: New York University Press.

EBCMGD (Editorial Board of Cha Ma Gu Dao), ed. 2003. *Cha Ma Gu Dao* (The ancient routes of tea and horses). Xi'an: Shanxi Shifan Daxue Chubanshe.

EBMCA (Editorial Board of Mengla Country Annals), ed. 1994. *Mengla Xian Zhi* (Mengla County annals). Kunming: Yunnan Renmin Chubanshe.

EBSAA (Editorial Board of Simao Area Annals), ed. 1996. *Simao Diqu Zhi* (Annals of Simao area). In *Zhonghua Renmin Gongheguo Difangzhi Congshu* (Local Chronicles Series of People's Republic of China, 2 vols.). Kunming: Yunnan Minzu Chubanshe.

Etherington, Dan M., and Keith Forster. 1993. *Green Gold: The Political Economy of China's Post-1949 Tea Industry.* Hong Kong: Oxford University Press.

Evans, John C. 1992. *Tea in China: The History of China's National Drink.* New York: Greenwood Press.

Fang Guoyu. 2001. Puer cha (Puer tea). In *Fang Guoyu Wenji: Di Si Ji* (Collected works of Fang Guoyu, vol. 4), edited by Lin Chaomin, 426–430. Kunming: Yunnan Jiaoyu Chubanshe.

Farquhar, Judith. 2002. *Appetites: Food and Sex in Postsocialist China.* Durham, N.C.: Duke University Press.

Feeley-Harnik, Gillian. 1995. Religion and food: an anthropological perspective. *Journal of the American Academy of Religion* 63: 565–582.

Ferguson, Priscilla Parkhurst. 1998. A cultural field in the making: gastronomy in 19th-century France. *The American Journal of Sociology* 104 (3): 597–641.

Filmsea. 2003. Wu Wenguang jianli yu zuopin yi lan (An overview of Wu Wenguang's CV and works). *Filmsea Website.* http://www.filmsea.com.cn/newsreel/celebrity/200304160021.htm (accessed 12 October 2008).

Forbes, Andrew D. W. 1987. The 'Cin-Ho' (Yunnanese Chinese) caravan trade with north Thailand during the late nineteenth and early twentieth centuries. *Journal of Asian History* 27: 1–47.

Fotoe. 2006. Yuanshengtai gewu ji Yunnan Yingxiang (Original song and dance performance Dynamic Yunnan). *Fotoe Website.* http://www.fotoe.com/sub/100973 (accessed 10 December 2006).

Friedberg, Suzanne. 2004. *French Beans and Food Scares: Culture and Commerce in an Anxious Age.* Oxford: Oxford University Press.

Gabaccia, Donna R. 1998. *We Are What We Eat: Ethnic Food and the Making of Americans.* Cambridge, Mass.: Harvard University Press.

Gao Fachang. 2009. *Gu Liu Da Cha Shan Shi Kao* (Historical research on the Ancient Six Great Tea Mountains). Kunming: Yunnan Meishu Chubanshe.

Gardella, Robert. 1994. *Harvesting Mountains: Fujian and the China Tea Trade, 1757–1937.* Berkeley: University of California Press.

Germann-Molz, Jennie. 2004. Tasting an imagined Thailand: authenticity and culinary tourism in Thai restaurants. In *Culinary Tourism,* edited by Lucy M. Long, 53–75. Lexington: The University Press of Kentucky

Giddens, Anthony. 1979. *Central Problems in Social Theory: Action, Structure and Contradiction in Social Analysis.* London: Macmillan.

Giersch, C. Patterson. 2006. *Asian Borderlands: The Transformation of Qing China's Yunnan Frontier.* Cambridge, Mass.: Harvard University Press.

Gold, Thomas B. 1993. Go with your feelings: Hong Kong and Taiwan popular culture in Greater China. *The China Quarterly* 136: 907–925.

Goodwin, Jason. 1993. *The Gunpower Gardens.* London: Vintage.

Goody, Jack. 1982. *Cooking, Cuisine, and Class: A Study in Comparative Sociology.* Cambridge: Cambridge University Press.

Grasseni, Cristina. 2003. Packaging skills: calibrating cheese to the global market. In *Commodifying Everything: Relationships of the Market,* edited by Susan Strasser, 259–288. New York: Routledge.

Grimshaw, Anna, and Nikos Papastergiadis. 1995. *Conversations with Anthropological Film-Maker David MacDougall.* Cambridge: Prickly Pear Press.

Guan Jianping. 2001. *Cha yu Zhongguo Wenhua* (Tea and Chinese culture). Beijing: Renmin Chubanshe.

Guo Yukuan. 2007a. Puer luan ju (The chaotic situation of Puer tea). *Xinshengdai* (New generations) 63: 14–40.

———. 2007b. Zou Jiaju: guanyu Puer cha de liu ge wujizhitan (Zou Jiaju: six groundless utterances about Puer tea). *Xinshengdai* (New generations) 63: 37–39.

Gupta, Akhil, and James Ferguson, eds. 1997. *Culture, Power, Place: Explorations in Critical Anthropology.* Durham, N.C.: Duke University Press.

Guy, Kolleen M. 2003. *When Champagne Became French: Wine and the Making of a National Identity.* Baltimore: Johns Hopkins University Press.

Hamm, John Christopher. 2005. *Paper Swordsmen: Jin Yong and the Modern Chinese Martial Arts Novel.* Honolulu: University of Hawaii Press.

Handler, Richard. 1986. Authenticity. *Anthropology Today* 2 (1): 2–4.

Harvey, David. 1989. *The Condition of Postmodernity: An Inquiry into the Origins of Cultural Change.* Cambridge: Basil Blackwell.

Haverluk, Terrence W. 2002. Chile peppers and identity construction in Pueblo, Colorado. *Journal for the Study of Food and Society* 6 (1): 45–59.

Hawkes, David, trans. 1957. *Ch'u Tz'u: The Songs of the South.* Oxford: Oxford University Press.

He Jingcheng. 2002. Cong Xianggang shichang jiaodu kan Puer cha fenlei (Puer tea categories in Hong Kong market). In *2002 Zhongguo Puer Cha Guoji Xueshu Yantaohui Lunwenji* (Proceedings of the 2002 International Academic Symposium on China Puer Tea), edited by Su Fanghua, 118–125. Kunming: Yunnan Renmin Chubanshe.

Heldke, L. 2005. But is it authentic? culinary travel and the search for the 'genuine article.' In *The Taste Culture Reader: Experiencing Food and Drink,* edited by C. Korsmeyer, 385–394. Oxford: Berg.

Hill, Ann Maxwell. 1989. Chinese dominance of the Xishuangbanna tea trade: an interregional perspective. *Modern China* 15 (3): 321–345.

———. 1998. *Merchants and Migrants: Ethnicity and Trade among Yunnanese Chinese in Southeast Asia.* New Haven, Conn.: Yale University Southeast Asia Studies.

Hillman, Ben. 2003. Paradise under construction: minorities, myths and modernity in northwest Yunnan. *Asian Ethnicity* 4 (2): 175–188.

Hilton, James. 1939. *Lost Horizon.* New York: William Morrow.

Ho Chi-p'eng. 1995. The lament for autumn: a type of time-space consciousness in the tradition of Chinese literature. Translated by Ch'en Chao-ying. In *Time and Space in Chinese Culture,* edited by Huang Chun-chieh and Erik Zürcher, 343–361. Leiden: E.J. Brill.

Hobsbawm, Eric. 1983. Introduction: inventing traditions. In *The Invention of Tradition,* edited by Eric Hobsbawm and Terence Ranger, 1–14. Cambridge: Cambridge University Press.

Hollander, Gail M. 2003. Re-naturalizing sugar: narratives of place, production and consumption. *Social and Cultural Geography* 4 (1): 59–74.

Hsing You-Tien. 2006. Comment on Zhang Li's 'Contesting spatial modernity in late-socialist China.' *Current Anthropology* 47 (3): 478.

Hsü Ching-wen. 2005. *Consuming Taiwan.* PhD diss., University of Washington.

Huang Anxi. 2004. *Le Yin Siji Cha* (Happily drink tea in four seasons). Translated by Sun Xiaoyan. Beijing: Shenghuo Dushu Xinzhi Sanlian Shudian.

Huang Guishu. 2002. Lun Yunnan Sipu qu gudai Puren dui zuguo cha wenhua fazhang de gongxian (The contribution of ancient Pu ethnic in Sipu Yunnan to the tea culture of China). In *2002 Zhongguo Puer Cha Guoji Xueshu Yantaohui Lunwenji* (Proceedings of the 2002 International Academic Symposium on China Puer Tea), edited by Su Fanghua, 110–117. Kunming: Yunnan Renmin Chubanshe.

———. 2005. *Puer Cha Wenhua Daguan* (A full display of Puer tea culture). Kunming: Yunnan Minzu Chubanshe.

Huang Yan, and Yang Zhijian. 2007. Wang zhe guilai: cong gugong dao Puer (The king is coming back: from Palace Museum to Puer). *Pu-Erh Special Issue* (April): 14–17.

Ismail, Mohamed Yusoff. 2002. Sacred food from the ancestors: edible bird nest harvesting among the Idaham. In *The Globalization of Chinese Food,* edited by David Y. H. Wu and Sidney C. H. Cheung, 43–55. Honolulu: University of Hawaii Press.

Ivy, Marilyn. 1995. *Discourses of the Vanishing: Modernity, Phantasm, Japan.* Chicago: University of Chicago Press.

Jameson, F. 1983. Nostalgia for the present. *South Atlantic Quarterly* 88 (2): 517–537.

Jiang Quan. (1980) 2006. Gu liu da cha shan fangwen ji (Record of visiting the Ancient Six Great Tea Mountains). In *Puer Cha Jingdian Wenxuan* (Classic anthology on Puer tea), edited by Wang Meijin, 33–47. Kunming: Yunnan Meishu Chubanshe.

Jing Wendong. 2003. *Liumang Shijie de Dansheng: Chong Du Jin Yong* (The birth of the hoodlum world: To reread Jin Yong). Guangzhou: Huacheng Chubanshe.

Jones, Mark, Paul Craddock, and Nicolas Barker, eds. 1990. *Fake? The Art of Deception.* Berkeley: University of California Press.

Kipnis, Andrew. 1995. 'Face': an adaptable discourse of social surfaces. *Positions: East Asia Cultures Critique* 3 (1): 119–148.

———. 1997. *Producing Guanxi: Sentiment, Self, and Subculture in a North China Village.* Durham, N. Carolina: Duke University Press.

———. 2006. Suzhi: a keyword approach. *China Quarterly* 186: 295–313.

Kong Chuizhu. 2007. Quanmian tuijin yi Puer cha wei daibiao de cha chanye maishang xin taijie: zai quansheng chaye gongzuo zuotan hui shang de jianghua (To fully promote the [Puer] tea industry development: a speech at the provincial tea working forum). *Puer Tea Weekly* (July): 55.

Kopytoff, Igor. 1986. The cultural biography of things. In *The Social Life of Things: Commodities in Cultural Perspective,* edited by Arjun Appadurai, 64–91. Cambridge: Cambridge University Press.

Kyllo, Jeffrey Alexander. 2007. Sichuan Pepper: The Roles of a Spice in the Changing Political Economy of China's Sichuan Province. BA thesis, University of Washington.

Lao Tzu. 1998. *Tao Te Ching.* Translated by Arthur Waley. Beijing: Waiyu Jiaoxue yu Yanjiu Chubanshe.

Latham, Kevin, Stuart Thompson, and Jakob Klein, eds. 2006. *Consuming China: Approaches to Cultural Change in Contemporary China.* London: Routledge.

Leach, Edmund. 1970. *Claude Lévi-Strauss.* New York: Viking.

Lei Pingyang. 2000. *Puer Cha Ji* (The biography of Puer tea). Kunming: Yunnan Minzu Chubanshe.

Leitch, Alison. 2003. Slow food and the politics of pork fat: Italian food and European identity. *Ethnos* 68 (4): 437–462.

Lévi-Strauss, Claude. 1970. *The Raw and the Cooked.* London: Jonathon Cape.

———. 2008. The culinary triangle. In *Food and Culture: A Reader,* edited by Carole Counihan and Penny Van Esterik, 36–44. New York: Routledge.

Li Fuyi. (1939) 2000. Fohai chaye gaikuang (The general situation of Fohai's tea industry). In *Puer Cha Ji* (The biography of Puer tea), edited by Lei Pingyang, 57–71. Kunming: Yunnan Minzu Chubanshe.

———. 1984. *Zhenyue Xian Xin Zhi Gao* (New chorography of Zhenyue County). Taipei: Furen Shu Wu.

Li Miao. 2006. Zhongguo Dianying Zhong de Yunnan Xiangxiang (The imagination of Yunnan in Chinese film). MA thesis, Shanghai University.

Li Quanmin. 2008. Identity, relationships and difference: the social life of tea in a group of Mon-Khmer speaking people along the China-Burma frontier. PhD diss., Australian National University.

Li Shi. 1988. Xu bo wu zhi (Sequel of natural history). In *Banna Wenshi Ziliao Xuanji Di Si Ji* (Selected works of historical accounts of Xishuangbanna, vol. 4), edited by Zhao Chunzhou and Zhang Shungao, 1. Kunming: Zhongguo Renmin Zhengzhi Xieshang Huiyi Xishuangbanna Daizu Zizhizhou Weiyuanhui Wenshi Ziliao Weiyuanhui. First published in the 12th century.

Li Yan, and Yang Zejun, eds. 2004. *Tianxia Puer* (Puer tea under heaven). Kunming: Yunnan Daxue Chubanshe.

Lien, Marianne E., and Brigitte Nerlich. 2004. *The Politics of Food.* Oxford: Berg.

Lin Chaomin. 2006. Puer cha san lun (A casual talk on Puer tea). In *Puer Cha Jingdian*

Wenxuan (Classic anthology on Puer tea), edited by Wang Meijin, 48–68. Kunming: Yunnan Meishu Chubanshe.

Liu, James J. Y. 1967. *The Chinese Knight-errant.* London: Routledge and Kegan Paul.

Liu Minjiang. 1983. Yiwu shangye ziben de tedian (Characteristics of Yiwu commercial capitalism). In *Daizu Shehui Lishi Diaocha. Xishuangbanna Zhi Yi* (Investigations of Xishuangbanna Dai Society, vol. 1), edited by Yunnan Compilation Committee, 57–61. Kunming: Yunnan Minzu Chubanshe.

Liu Qinjin. 2005. *Zhongguo Puer Cha Zhi Kexue Du Ben* (A scientific book for China's Puer tea). Guangzhou: Guangdong Lüyou Chubanshe.

Liu Yanwu, ed. 2003. *Zhongguo Jianghu Yinyu Cidian* (Dictionary for Chinese *jianghu* slang). Beijing: Zhongguo Shehui Kexue Chubanshe.

Lowenthal, David. 1985. *The Past Is a Foreign Country.* Cambridge: Cambridge University Press.

Lu Ming. 2007. Jingti gei Puer cha mohei de shou (A precaution for those who slander Puer tea). *Evening Newspaper of Spring City,* 9 July.

Lu Yu. 2003. *Cha Jing* (The classic of tea). Beijing: Zhongguo Gongren Chubanshe. First published in the 8th century.

Luo Qun. 2004. *Jindai Yunnan Shangren yu Shangren Ziben* (Yunnan merchants and merchant capital in modern times). Kunming: Yunnan Daxue Chubanshe.

Lupton, Deborah. 1996. *Food, the Body and the Self.* London: Sage Publications.

Ma Jianxiong. 2007. Ailao Shan fudi de zuqun zhengzhi: Qing zhong qian qi 'gaitu guiliu' yu 'Luohei' de xingqi (The ethnic politics in the hinterland of Ailao Mountain: gaitu guiliu and the rise of Luohei in the early and mid Qing). *Zhongyang Yanjiuyuan Lishi Yuyan Yanjiusuo Jikan* (Collected papers of the Research Institute of History and Language, Academia Sinica) 78 (3): 553–600.

Ma Yihua. 2006. Puer cha: ceng yiluo taxiang de mingpian (Puer tea: a name card lost in an alien land). *Yunnan Daily Website,* http://paper.yunnan.cn/html/20060622/news_89_285115.html (accessed 8 August 2006).

Macfarlane, Alan, and Iris Macfarlane. 2003. *Green Gold: The Empire of Tea.* London: Ebury Press.

Mair, Victor H., and Erling Hoh. 2009. *The True History of Tea.* London: Thames and Hudson.

Maule, Robert. 1991. Tea production on the periphery of the British Empire. *Thai-Yunnan Project Newsletter* 14 (September). http://www.nectec.or.th/pub/info/thai-yunnan/thai-yunnan-nwsltr-14.txt. (accessed 24 November 2008).

Mauss, Marcel. 1954. *The Gift: Forms and Functions of Exchange in Archaic Societies.* London: Routledge and Kegan Paul.

May, Rachel, and John Minford. 2004. General glossary of terms. In *The Book and the Sword: A Martial Arts Novel,* by Jin Yong, translated by Grahmand Earnshaw, edited by Rachel May and John Minford, xvii-xxii. New York: Oxford University Press.

Mengla Archive. 1982. Yiwu dadui gai wei cha nong de youguan guiding (Relevant regulations on changing Yiwu team into tea peasants). In *Mengla Archive Document,* Mengla County.

Menzies, Nick. 2008. Forest tea and terrace tea in Xishuangbanna: tradition, ethnicity, and science on the road to modernity. Paper presented at the Thai Studies Conference, Thammasat University, Bangkok, January 9–11 2008.

Messer, Ellen 1984. Anthropological perspectives on diet. *Annual Review of Anthropology* 13: 205–249.

Miller, Daniel. 1997. *Material Cultures: Why Some Things Matter.* Chicago: University of Chicago Press.

Minford, John. 1997. General glossary of terms. In *The Deer and the Cauldron: A Martial Arts Novel,* by Jin Yong, translated and edited by John Minford, xxv-xxxi. New York: Oxford University Press.

Mintz, Sidney Wilfred. 1985. *Sweetness and Power: The Place of Sugar in Modern History.* New York: Viking.

———. 1996. *Tasting Food, Tasting Freedom: Excursions into Eating, Culture, and the Past.* Boston: Beacon Press.

Mu Jihong. 1992. *Dian Chuan Zang Da Sanjiao Wenhua Tan Mi* (Cultural search among the triangle of Yunnan, Tibet, and Sichuan). Kunming: Yunnan Daxue Chubanshe.

———. 2003. *Cha Ma Gu Dao Shang de Minzu Wenhua* (The ethnic culture along the ancient routes of tea and horses). Kunming: Yunnan Minzu Chubanshe.

———. 2004. *Puer Cha Shuo* (The book of Puer tea). Beijing: Zhongguo Qing Gongye Chubanshe.

National Palace Museum (NPM). 2010. Painting and calligraphy of the Northern Sung (960–1127). *National Palace Museum Website.* http://tech2.npm.gov.tw/sung/html/graphic/c_t3_1_b21.htm (accessed 6 June 2010).

NBQIQC (National Bureau of Quality Inspection and Quarantine of China). 2008. Guanyu pizhun dui Puer cha shishi dili biaozhi chanpin baofu de gonggao (The announcement confirming Puer tea as a geographical product). *Pu-Erh* 15 (December).

NBSC (National Bureau of Statistics of China). 2009. GDP growth in China, 1952–2009. *Chinability Website.* http://www.chinability.com/GDP.htm (accessed 4 April 2010).

Ni Tui. (1737) 1981. Dian yun li nian zhuan, juan shi'er (Document of Yunnan history, vol. 12). In *Zhongguo Chaye Lishi Ziliao Xuanji* (Selected historical documents on Chinese tea), edited by Chen Zugui and Zhu Zizhen, 593–594. Zhongguo Nong Shi Zhuanti Ziliao Huibian (Special collection on Chinese agriculture). Beijing: Nongye Chubanshe.

Notar, Beth E. 2006a. Authenticity anxiety and counterfeit confidence: outsourcing souvenirs, changing money, and narrating value in Reform-era China. *Modern China* 32 (1): 64–98.

———. 2006b. *Displacing Desire: Travel and Popular Culture in China.* Honolulu: University of Hawaii Press.

Oakes, Timothy S. 1997. Ethnic tourism in rural Guizhou: sense of place and the commerce of authenticity. In *Tourism, Ethnicity, and the State in Asian and Pacific Societies,* edited by M. Picard and R. Wood, 35–70. Honolulu: University of Hawaii Press.

Oakes, Timothy S., and Louisa Schein. 2006. Translocal China: an introduction. In *Translocal China: Linkages, Identities, and the Reimagining of Space,* edited by Tim Oakes and Louisa Schein, 1–35. London: Routledge.

Ohnuki-Tierney, Emiko. 1990. The ambivalent self of the contemporary Japanese. *Cultural Anthropology* 5 (2): 197–216.

———. 1993. *Rice as Self: Japanese Identities through Time.* Princeton, N.J.: Princeton University Press.

Ortner, Sherry B. 2006. *Anthropology and Social Theory: Culture, Power, and the Acting Subject.* Durham, N.C.: Duke University Press.

Ozeki, Erino. 2008. Fermented soybean products and Japanese standard taste. In *The World of Soy,* edited by Christine M. Du Bois, Chee-Beng Tan, and Sidney W. Mintz, 144–160. Singapore: NUS Press.

Perdue, Peter C. 2008. Is Puer in Zomia? Tea cultivation and the state in China. Paper presented at the Agrarian Studies Colloquium, Yale University, October 24.

Peters, John Durham. 1997. Seeing bifocally: media, place, culture. In *Culture, Power, Place: Explorations in Critical Anthropology,* edited by Akhil Gupta and James Ferguson, 75–92. Durham, N.C.: Duke University Press.

Phillips, Lynne. 2006. Food and globalization. *Annual Review of Anthropology* 35: 37–57.

Prasertkul, Chiranan. 1989. *Yunnan Trade in the Nineteenth Century: Southwest China's Cross-Boundaries Functional System.* Bangkok: Institute of Asian Studies, Chulalongkorn University.

Pu-Erh. 2007a. Beihuiguixian shang de huigui (Returning at the Tropic of Cancer). Special Issue, *Pu-Erh* (April): 4–5.

———. 2007b. Puer da shi ji (Record of the major events in Puer). Special Issue, *Pu-Erh* (April): 43.

Puer Jianghu. 2007a. Kai dian re (Fever of opening tea shops). *Puer Jianghu* (June).

———. 2007b. Puer jianghu jiaotou (The coach of the Puer *jianghu*). *Puer Jianghu* (April).

Puer Tea Weekly. 2007a. Laizi Yunnan de shengyin: Puer cha xianzhuang zhi zheng ben qing yuan daxing zhuanti zuotanhui zai kun zhaokai (The voices from Yunnan: a symposium was held in Kunming to clarify the current situation of Puer tea). *Puer Tea Weekly* 53 (June–July).

———. 2007b. 'Zheng ben qing yuan shihua shishuo Puer cha' huodong zhengshi qidong (Launching the activity of clarifying and telling the truth about Puer tea). *Puer Tea Weekly* 65–72 (September–November).

———. 2007c. Zhongguo 'chaye gainian gu' beihou de ziben youxi (The capital game behind Chinese 'tea conceptual stock'). *Puer Tea Weekly* 59 (August).

Riftea. 2007. 2005 Yunnan mabang rui gong jingcheng huodong (2005 Yunnan caravan tribute event). *Ruijingfang Tea Website* http://www.rjftea.com/rjf/shownews.asp?id=48&BigClass=%D2%B5%C4%DA%D7%CA%D1%B6 (accessed 6 June 2007).

Ruan Dianrong. 2005a. *Liu Da Cha Shan* (The six great tea mountains). Beijing: Zhongguo Qing Gongye Chubanshe.

———. 2005b. *Wo de Renwen Puer* (My humanity Puer tea). Kunming: Yunnan Renmin Chubanshe.

Ruan Fu. (1825) 1981. Puer cha ji (Biography of Puer tea). In *Zhongguo Chaye Lishi Ziliao Xuanji* (Selected historical documents on Chinese tea), edited by Chen Zugui and Zhu Zizhen, 396–397. Beijing: Nongye Chubanshe.

Sanlian Life Week (Sanlian shenghuo zhoukan). 2012. Zhuanfang Xu Ke (An interview with Xu Ke). *Sanlian Life Week* (2): 62.

Sanzui. 2007. 2007 Gudao Puer chencha Kunming hui hou pin hou gan (Taste comments of ancient road and aged tea tasting event in Kunming 2007). *Sanzui Website.* http://www.sanzui.com/bbs/viewthread.php?tid=75261 (accessed 12 December 2007).

Schein, Louisa. 2000. *Minority Rules: The Miao and the Feminine in China's Cultural Politics.* Durham, N.C.: Duke University Press.

Scott, James C. 1985. *Weapons of the Weak: Everyday Forms of Peasant Resistance.* New Haven: Yale University Press.

———. 1998. *Seeing Like a State: How Certain Schemes to Improve the Human Condition Have Failed.* New Haven: Yale University Press.

———. 2009. *The Art of Not Being Governed: An Anarchist History of Upland Southeast Asia.* New Haven: Yale University Press.

Seremetakis, C. Nadia. 1994. *The Senses Still: Perception and Memory as Material Culture in Modernity.* Boulder, Colo.: Westview Press.

Shao Wanfang. 2007. Qinggong gongcha pinyin ji (Record of tasting tribute tea of Qing legacy). *Pu-Erh* 3 (6): 100–103.

Shen Dongmei. 2007. *Cha yu Songdai Shehui Shenghuo* (Tea and social life during the Song dynasty). Beijing: Zhongguo Shehui Kexue Chubanshe.

Shen Peiping. 2007. Yi gengming wei qiji, puxie Puer fazhan xin pianzhang (Taking name change as an opportunity to compose a new chapter for Puer's development). *Yunnan Daily,* 6 April.

———, ed. 2008. *Zoujin Chashu Wangguo* (Entering the kingdom of tea trees). Kunming: Yunnan Keji Chubanshe.

Shengse Chama. 2007. Huyou mianyu: zhongshi yu ganjue benshen (To immunize against *huyou:* be loyal to your own senses). *Sanzui Website* http://www.sanzui.com/bbs/showthread.php?t=75049 (accessed 30 July 2007).

Shi Junchao, ed. 1999. *Hani Zu Wenhua Daguan* (A full display of the Hani culture). Kunming: Yunnan Minzu Chubanshe.

Shi Kunmu. 2005. *Jingdian Puer* (The scripture of Puer). Beijing: Tong Xin Chubanshe.

Sima Qian. 2011. *Shi Ji* (Historical records). Shanghai: Shanghai Guji Chubanshe. First published in the 10th century B.C.E.

Sterckx, Roel, ed. 2005. *Of Tripod and Palate: Food, Politics and Religion in Traditional China.* New York: Palgrave Macmillan.

Strasser, Susan, ed. 2003. *Commodifying Everything: Relationships of the Market.* New York: Routledge.

Su Fanghua. 2002. Hongyang Puer cha wenhua rang gengduo de ren xiai Puer cha (Promoting Puer tea culture to have more Puer tea lovers). In *2002 Zhongguo Puer Cha Guoji Xueshu Yantaohui Lunwenji* (Proceedings of the 2002 International Academic Symposium on China Puer Tea), edited by Su Fanghua, 48–58. Kunming: Yunnan Renmin Chubanshe.

Su Heng-an. 2004. *Culinary Arts in Late Ming China: Refinement, Secularization and Nourishment.* Taipei: SMC Publishing.

Sutton, David. 2001. *Remembrance of Repasts: An Anthropology of Food and Memory.* Oxford: Berg.

Tagliacozzo, Eric, and Wen-chin Chang, eds. 2011. *Chinese Circulations: Capital, Commodities and Newworks in Southeast Asia.* Durham, N.C.: Duke University Press.

Talbot, John M. 2004. *Grounds for Agreement: the Political Economy of the Coffee Commodity Chain.* Lanham, Md.: Rowman and Littlefield.

Tam, Siumi Maria. 2002. Heunggongyan forever: immigrant life and Hong Kong style Yumcha in Australia. In *The Globalization of Chinese Food,* edited by David Y. H. Wu and Sidney C. H. Cheung, 131–151. Honolulu: University of Hawaii Press.

Tan, Chee-beng, and Ding Yulin. 2010. The promotion of tea in south China: re-inventing tradition in an old industry. *Food and Foodways* 18 (3): 121–144.

Tan Cui. (1799) 1981. Dian hai yu heng zhi. In *Zhongguo Chaye Lishi Ziliao Xuanji* (Selected historical documents on Chinese tea), edited by Chen Zugui and Zhu Zizhen, 387. Beijing: Nongye Chubanshe.

Tang Jianguang, Huan Li, and Wang Xun. 2007a. Puer de shengshi weiyan (A cautious speech on Puer tea's heyday). *China Newsweek* 325 (19): 26–34.

———. 2007b. Zhengfu de juese (The role of the government). *China Newsweek* 325 (19): 30–33.

Tapp, Nicholas. 2003. Exiles and reunion: nostalgia among overseas Hmong (Miao). In *Living with Separation in China: Anthropological Accounts,* edited by Charles Stafford, 57–75. London: Routledge and Curzon.

Terrio, Susan J. 2005. Crafting grand cru chocolates in contemporary France. In *The Cultural Politics of Food and Eating: A Reader,* edited by James L. Watson and Melissa L. Caldwell, 144–162. Malden, Mass.: Blackwell.

Toomey, Paul Michael, ed. 1994. *Food from the Mouth of Krishna: Feasts and Festivals in a North Indian Pilgrimage Centre.* Delhi: Hindustan Publishing Corp.

Trevor-Roper, Hugh. 1983. The invention of tradition: the highland tradition of Scotland. In *The Invention of Tradition,* edited by Eric Hobsbawm and Terence Ranger, 15–42. Cambridge: Cambridge University Press.

Trilling, Lionel. 1974. *Sincerity and Authenticity.* London: Oxford University Press.

Ukers, William H. 1935. *All about Tea.* 2 vols. New York: Tea and Coffee Trade Journal Co.

Ulin, Robert C. 1996. *Vintages and Traditions: An Ethnohistory of Southwest French Wine Cooperatives.* Washington, D.C.: Smithsonian Institution Press.

Wang Jing. 1996. *High Culture Fever: Politics, Aesthetics, and Ideology in Deng's China.* Berkeley: University of California Press.

Wang Xun. 2007. Puer cha shifo shenqi? (Is Puer tea miraculous?). *China Newsweek* 325 (19): 31.

Watson, James L. 1997. *Golden Arches East: McDonald's in East Asia.* Stanford, Calif.: Stanford University Press.

Wilk, Richard, eds. 2006. *Fast Food/Slow Food: The Cultural Economy of the Global Food System.* Lanham, Md.: Altamira Press.

Wu, David Y. H., and Sidney C. H. Cheung, eds. 2002. *The Globalization of Chinese Food.* Honolulu: University of Hawaii Press.

Wu, David Y. H., and Tan Chee Beng. 2001. *Changing Chinese Foodways in Asia.* Hong Kong: Chinese University Press.

Wu Qiong. 2006. Yunnan Puer cha wenhua Zhongguo ming shan xing huodong qidong (The launch of storing Yunnan Puer tea in the famous mountains of China). *Yunnan Puer Cha* (Spring): 131.

Wu Wenguang. 1999. *Jianghu.* 150 minutes.

Xie Zhaozhi. 2005. Dian lue: juan san (Yunnan account, vol. 3). In *Puer Cha Jingdian Wenxuan* (Classic anthology on Puer tea), edited by Wang Meijin, 3. Kunming: Yunnan Meishu Chubanshe. First published in the early 17th century.

Xinjing. 2006. Wanli cha ma gu dao shi yue zoujin Nibo'er (Long journey of caravan arrived in Nepal in October). *Xinjing Website.* http://www.5caishi.com/Get/yjnews/08181566.htm (accessed 10 September 2006).

Xu Jianchu. 2007. Rattan and tea-based intensification of shifting cultivation by Hani farmers in southwestern China. In *Voices from the Forest: Integrating Indigenous Knowledge into Sustainable Upland Farming,* edited by Malcolm Cairns, 667–675. Washington, D.C.: Resources for the Future.

Xu Yahe. 2006. *Jie Du Puer: Zui Xin Puer Cha Bai Wen Bai Da* (Interpreting Puer tea: The newest one hundred questions and answers on Puer tea). Kunming: Yunnan Meishu Chubanshe.

Yan Yunxiang. 1996. *The Flow of Gifts: Reciprocity and Social Networks in a Chinese Village.* Stanford, Calif.: Stanford University Press.

———. 2009. *The Individualization of Chinese Society.* Oxford: Berg.

Yang Bin. 2004. Horses, silver, and cowries: Yunnan in global perspective. *Journal of World History* 15 (3): 281–322.

———. 2006. *Between Winds and Clouds: The Making of Yunnan (Second Century B.C.E. to Twentieth Century C.E.).* Gutenberg-e. http://www.gutenberg-e.org/yang/acknowledgments.html (accessed 6 February 2010).

Yang Haichao. 2007. Yunnan cha shi kao wu (Research on Yunnan tea's history). In *Puer Cha Wenhua Cidian* (A glossary of Puer tea culture), edited by Mu Jihong, 27–59. Beijing: Jixie Gongye Chubanshe.

———. 2010. Cha wenhua chu chuan zang qu de shijian yu kongjian (The initial time and space of tea arriving in Tibet). *Qinghai Minzu Yanjiu* (Studies of Qinghai's ethnic groups) (3): 111–115.

Yang Kai, Liu Yan, and Li Xiaomei. 2008. *Cong Da Qing Dao Zhongcha: Zui Zhenshi de Puer Cha* (From the Qing dynasty to Zhongcha: The most authentic stories about Puer tea). Kunming: Yunnan Renmin Chubanshe.

Yang, Mayfair. 1988. The modernity of power in the Chinese socialist order. *Cultural Anthropology* 3 (4): 408–426.

———. 1989. The gift economy and state power in China. *Comparative Studies in Society and History* 31 (1): 25–54.

———. 1994. *Gifts, Favors, and Banquets: The Art of Social Relationships in China.* Ithaca, N.Y.: Cornell University Press.

YRTN (*Yunnan Radio and Television Newspaper*). 2005. The revival of Yunnan films. *Yunnan Radio and Television Newspaper,* 23 February.

YTG (Yiwu Township Government). 2007. *Yiwu Gu Zhen Jianshe he Cha Chanye Fazhan de Silu he Cuoshi* (Thoughts and measures on construction of the ancient town Yiwu and development of its tea industry). Yiwu government document.

YTIEC (Yunnan Tea Import and Export Corporation). 1993. *Yunnan Tea Import and Export Corporation Annals, 1938–1990.* Kunming: Yunnan Renmin Chubanshe.

Yu Shuenn-der. 2010. Materiality, stimulants and the Puer tea fad. *Journal of Chinese Dietary Culture* 6 (1): 107–142.

Yuan Mei. 1792. *Suiyuan Shidan (Di Liu Jie)* (Recipes from the Sui garden, chapter 6). Website of Art China, http://guji.artx.cn/article/9465.html (accessed 14 May 2010).

Yunnan Daily. 2006a. Cha shi zang bao xinzhong de Yunnan yinxiang (Tea represents Yunnan for Tibetan people). *Yunnan Daily Website.* http://www.5caishi.com/Get/yjnews/08181566.htm (accessed 10 September 2006).

———. 2006b. Gedebao Hao yu Puer cha (*Göthenburg* and Puer tea). *Yunnan Daily Website.* http://www.yndaily.com/ihtml/yndaily/TXTPA_GDB06.html (accessed 2 February 2007).

———. 2006c. Kunming: Puer cha zai chuang huihuang de jisan di (Kunming: the distribution center of promoting Puer tea). *Yunnan Daily Website.* http://paper.yunnan.cn/html/20061209/news_92_33061.html (accessed 10 October 2008).

Yunnan Puer Cha. 2006. Yunnan Puer cha jiang cheng Gedebao hao chongfan Ouzhou dalu (Yunnan Puer tea will board *Göthenburg* to return to Europe). *Yunnan Puer Cha* (Spring): 136–137.

———. 2007. Puer jingji: cui sheng duo zhong Puer cha meiti (Puer tea economy boosted multiple Puer tea media). *Yunnan Puer Cha* (Winter): 39–41.

Zeng Zhixian. 2001. *Fangyuan Zhi Yuan: Shen Tan Jin Ya Cha Shijie* (Experience within the micro circumference: A deep exploration in the world of compressed tea). Taipei: Zeng Zhixian.

Zhang Hong. 1998. Dian nan xin yu (New narratives about southern Yunnan). In *Zhongguo Chaye Lishi Ziliao Xuanji* (Selected historical documents on Chinese tea), edited by Chen Zugui and Zhu Zizhen, 369. Beijing: Nongye Chubanshe. First published c. 1755.

Zhang Jinghong. 2010a. Remorse about the "Authentic Mountain Tea": packaging Puer tea in Yiwu. *Journal of Chinese Dietary Culture* 6 (2): 103–144.

———. 2010b. Multiple visions of authenticity: Puer tea consumption in Yunnan and other places. *Journal of Chinese Dietary Culture* 6 (1): 63–106.

———. 2012a. In between "the raw" and "the cooked": the cultural speculation and debate on Puer tea in contemporary China. *Harvard Asia Quarterly* Spring/Summer XIV (1 & 2): 44–52.

———. 2012b. Puer tea and rural transformation. In *Cha Ma Gu Dao Jikan* (Collected papers on the Ancient Tea-Horse Road, vol. 2), edited by Wang Shiyuan. 64–81. Kunming: Yunnan Daxue Chubanshe.

Zhang Li. 2006. Contesting spatial modernity in late-socialist China. *Current Anthropology* 47 (3): 461–484.

Zhang Shungao. 1988. Xishuangbanna chaye shengchan de guoqu, xianzai he weilai (The past, present, and future of Xishuangbanna's tea). In *Banna Wenshi Ziliao Xuanji Di Si Ji* (Selected works of historical accounts of Xishuangbanna, vol. 4), edited by Zhao Chunzhou and Zhang Shungao, 76–120. Kunming: Zhongguo Renmin Zhengzhi Xieshang Huiyi Xishuangbanna Daizu Zizhizhou Weiyuanhui Wenshi Ziliao Weiyuanhui.

Zhang Shungao, and Su Fanghua, eds. 2007. *Zhongguo Puer Cha Baikequanshu: Chanye Juan* (Encyclopedia of China's Puer tea: the industrial volume). Kunming: Yunnan Keji Chubanshe.

Zhang Yi. 2006a. *Gu Liu Da Cha Shan Jishi* (Document of the Ancient Six Great Tea Mountains). Kunming: Yunnan Minzu Chubanshe.

———. 2006b. Zhongguo Puer cha gu liu da chashan de guoqu he xianzai (The past and the present of the Ancient Six Great Mountains for China Puer tea). In *Puer Cha Jingdian Wenxuan* (Classic anthology on Puer tea), edited by Wang Meijin, 69–80. Kunming: Yunnan Meishu Chubanshe.

Zhang Yingpei. 2006. *Zhongguo Puer Cha Gu Liu Da Cha Shan* (The Ancient Six Great Tea Mountains of China Puer tea). Kunming: Yunnan Meishu Chubanshe.

———. 2007. Puer Cha Yuan Chan Di Xishuangbanna (Xishuangbanna: The original hometown of Puer tea). Kunming: Yunnan Keji Chubanshe.

Zhang Zhongliang and Mao Xianjie, eds. 2006. *Zhongguo Shijie Cha Wenhua* (Tea culture of China and the world). Beijing: Shishi Chubanshe.

Zhao Chunzhou and Zhang Shungao, eds. 1988. *Banna Wenshi Ziliao Xuanji Di Si Ji* (Selected works of historical accounts of Xishuangbanna, vol. 4). Kunming: Zhongguo Renmin Zhengzhi Xieshang Huiyi Xishuangbanna Daizu Zizhizhou Weiyuanhui Wenshi Ziliao Weiyuanhui.

Zhao Rubi, ed. 2006. *Li Lan Xishuangbanna Gu Cha Shan* (An overview of Xishuangbanna's ancient tea mountains). Kunming: Yunnan Minzu Chubanshe.

Zhao Zhichun. 1988. Puer fu zhi chaye ji jie (Overal annotation on the local tea history of Puer). In *Banna Wenshi Ziliao Xuanji Di Si Ji* (Selected works of historical accounts of Xishuangbanna, vol. 4), edited by Zhao Chunzhou and Zhang Shungao, 7–13. Kunming: Zhongguo Renmin Zhengzhi Xieshang Huiyi Xishuangbanna Daizu Zizhizhou Weiyuanhui Wenshi Ziliao Weiyuanhui.

Zheng Yongjun. 2007. Bainian huigui de zhuangju: Puer shiwei fushuji Zhu Yunfei fangtan (The spectacle of the one-hundred-year-old's return: talk to the vice secretary of Puer Municipal Party Committee Zhu Yunfei). *Pu-Erh Special Issue* (April): 8–9.

Zhou Hongjie, ed. 2004. *Yunnan Puer Cha* (Yunnan Puer tea). Kunming: Yunnan Keji Chubanshe.

———, ed. 2007. *Puer Cha Jiankang Zhi Dao* (Puer tea's health benefits). Xi'an: Shanxi Renmin Chubanshe.

Zhu Sikun, and Li Yin. 2006. Caiyun zhi nan de wenhua pai: zhuanfang Yunnan shengwei fushuji Dan Zeng (The culture brand beyond the cloud: an interview with Dan Zeng, the price clerk of the Communist Party of Yunnan). *China Report* 9: 14–19.

Zhu Xiaohua. 2007. Man de yisi (The meaning of slowness). *Puer Jianghu,* May.

Zhu Zizhen, ed. 1996. *Cha Shi Chu Tan* (Initial research on the history of tea). Beijing: Zhongguo Nongye Chubanshe.

Zou Jiaju. 2004. *Man Hua Puer Cha* (Casual talk about Puer tea). Kunming: Yunnan Minzu Chubanshe.

———. 2005. *Man Hua Puer Cha: Jingetiema Da Ye Cha* (Casual talk about Puer tea: large-leaf tea during wars). Kunming: Yunnan Meishu Chubanshe.

INDEX